Palgrave Studies in Creativity and Culture

Both creativity and culture are areas that have experienced a rapid growth in interest in recent years. Moreover, there is a growing interest today in understanding creativity as a socio-cultural phenomenon and culture as a transformative, dynamic process. Creativity has traditionally been considered an exceptional quality that only a few people (truly) possess, a cognitive or personality trait 'residing' inside the mind of the creative individual. Conversely, culture has often been seen as 'outside' the person and described as a set of 'things' such as norms, beliefs, values, objects, and so on. The current literature shows a trend towards a different understanding, which recognises the psycho-socio-cultural nature of creative expression and the creative quality of appropriating and participating in culture. Our new, interdisciplinary series Palgrave Studies in Creativity and Culture intends to advance our knowledge of both creativity and cultural studies from the forefront of theory and research within the emerging cultural psychology of creativity, and the intersection between psychology, anthropology, sociology, education, business, and cultural studies. Palgrave Studies in Creativity and Culture is accepting proposals for monographs, Palgrave Pivots and edited collections that bring together creativity and culture. The series has a broader focus than simply the cultural approach to creativity, and is unified by a basic set of premises about creativity and cultural phenomena.

More information about this series at
http://www.palgrave.com/gp/series/14640

Juan A. Roche Cárcel

Creativity and Time: A Sociological Exploration

palgrave
macmillan

Juan A. Roche Cárcel
University of Alicante
Alicante, Spain

Palgrave Studies in Creativity and Culture
ISBN 978-3-030-84837-8 ISBN 978-3-030-84838-5 (eBook)
https://doi.org/10.1007/978-3-030-84838-5

Cover illustration: LionH/Getty

This Palgrave Macmillan imprint is published by the registered company Springer Nature Switzerland AG.
The registered company address is: Gewerbestrasse 11, 6330 Cham, Switzerland

For Carmen Marimón Llorca, the source of my creativity

Contents

Part I

Introduction

Although for Sociology the study of creativity has traditionally occupied the margin and is an "elusive object and difficult to define," in recent times it has become an emerging research object, in particular since *The Creativity of Action* by Hans Joas (2013). Perhaps for this reason the sociological approach to this concept suffers from a certain imprecision, since it is frequently approached as something *a priori* and admitted without reflecting on what it means socially and culturally. On the other hand, in Sociology that investigates the subject of creativity, it is common to find the idea that it seeks novelty, innovation, and originality, but with little specificity.

With the aim of breaking such inconcretion, the author's perspective defends that the pursuit of originality constitutes one of the most important characteristics of creativity, but that originality refers, etymologically, to both origin and originary. Hence, the book is structured into two large sections, dedicated, respectively, to the creative categories of origin and the creative categories of originary. Among the former are creation myths, games—the origin of all cultural activity, the dialectic chaos–order, axial civilizations—the germ of our time, and the struggle between generations—a factor of social transformation. And, among the second, creative capitalism, creative work in the context of the global economy of risk and uncertainty, and representative democracy.

However, these two concepts are not isolated, but deeply interrelated, in a way that explains how creative originality builds a temporal narrative. Both merge the three temporalities—past, present, and future—and are constituted as a process fully immersed in time and in History, but in a dialectical way, since they are the makers and products of creativity. If society, in its long temporal duration, weaves a narrative fabric that inseparably unites it with time, undoubtedly creativity does so too. This is so because originality resolves—creatively—through a temporary process that is dialectical, the tension that exists between tradition and the avant-garde, continuity and discontinuity, and their social agents, the ancient and the modern. And, ultimately, because, by acting in this way, it tries to embrace memory and hope, critically recall the collective memory, and, at the same time, build new utopias. Consequently, creativity intensifies, vivifies, and rejuvenates the sense of a present located between the origin and the originary.

The book concludes that this temporal narrative has been dislocated in late modernity and, with it, creativity has been broken. First, because today it has neither space nor time; second, because it has gone from seeking fusion (characteristic of the categories of origin) to fragmentation and inequality (specific of the originary categories); and, finally, because creativity has been transformed from wishing for the survival of society, and from the existential flight of time, as happened in ancient myths, into a time-consuming creativity and consumed by it, in the aging originary categories. This process of transformation of creativity has followed two phases in modernity: the infantile and youthful rejuvenation of the first and the aging of the second. The latter has been caused, basically, by the role of the world wars, the demographic aging of society and the senescence of capitalism, contemporary art and science, and the crisis of representative democracy.

This generalized creative senility raises broad questions about the possibility that creativity, as Hans Joas suggests, may represent, at least in the short term, a way out of the contemporary crisis. In fact, it becomes a symptom of that crisis, rather than an opportunity for transformation. This book insists that the breakdown of the temporal narrative that has traditionally been brought about by creative originality and that connected the past, the present, and the future could be at the base of that transforming impossibility that creativity adopts today.

Juan A. Roche Cárcel, University of Alicante, June 2021.

1

Originality or from the Origin to the Originary

As will be seen in this chapter, the sociological approach to the concept of creativity lacks some accuracy, since it is addressed from an a priori perspective and admitted without reflecting on what it means socially and culturally. Habitually sociological literature that investigates creativity seeks novelty, originality. However, it is not usual for sociologists to specify what originality means and to link it to time. In this chapter the etymology of originality refers, dialectically, to the neighboring terms of "origin" and "originating." While the first supposes, basically, a return to the past and the second, a projection toward the future, creativity generates something original only when it develops a process, a temporal narrative, which expands the present and relates it, inseparably, with the past and the future.

1 Theoretical and Methodological Foundations

This book is supported on the following theoretical and methodological bases. First and foremost, the analysis essentially starts from a Weberian orientation of our discipline or, expressed differently, from Comprehensive

J. A. Roche Cárcel, *Creativity and Time: A Sociological Exploration*, Palgrave Studies in Creativity and Culture, https://doi.org/10.1007/978-3-030-84838-5_1

or Interpretive Sociology (Weber, 2006, pp. 13, 43–4 and 172; González García, 1992, p. 37, 1998, p. 208), according to which the social world and the relationships generated by it are meaningful. Thus, "comprehension" requires a "rational interpretation" (Weber, 1971, p. 150), whereas "signification" provides the data that the sociologist works with and which allows the latter to find, through the concepts of "correspondence in meaning" or "elective affinities," the common links between the different cognitive dimensions—aesthetic, ethical, economic, political, religious, and social—that modernity has fragmented and, accordingly, to restore the sense, the "worldview" (Muñoz, 2001, p. 23).

The aforementioned Weberian sociological theory is complemented with a methodology, the "heuristic or interpretive method" coming from Hermeneutic Philosophy, presented by Hans-Georg Gadamer—in *Truth and Method*, 1960—as a new philosophical way of thinking (Domínguez, 1997, p. 98). It is indeed grounded on a philosophical reflection which, in addition to having interpretation as its key concern (Ricoeur, 1988, p. 39), deals with the experience of understanding and the axial role played by the interpretation of the human sphere when it comes to its relationship with the world, in such a way that it is precisely in interpreting that the possibility to make sense of it lies.

It is of particular interest, in this respect, to apprehend the meanings that works of art convey; as a matter of fact, they constitute an experience of sense (Domínguez, 1997, pp. 94–98), a consummated response by means of which human existence understands itself (Gadamer, 1997, pp. 20–21). This is applicable especially in social life, since Philosophical Hermeneutics turns out to be highly useful for the knowledge of social phenomena, insofar as aesthetic and social facts make up an inseparable whole (Fernández and García, 2012–2013, p. 51). In that regard, modernity represents both a de-aesthetization and an aesthetization machinery because, inside it, creativity refers us back to the emergence of the new within a preferably lasting state and, simultaneously, to the aesthetic as an activity, without forgetting that it takes the artist as a model (Reckwitz and Sánchez Capdequí 2017, pp. 356–361). On the other hand, the unity between the aesthetic and the social spheres also becomes evident in the text, which is riddled with cultural connotations—in the sender's pole and in that of the receiver. For this reason, what we interpret are the things themselves, though seen in their own context. Therefore, it does not focus on language, but on a text (Lince, 2009, 19 ff.)—the fundamental basis of interpretation (Beltrán, 2012, p. 21)—which never has an autonomous nature, since it is contextualized (Beltrán, 2016, pp. 3–4), such contextualization precisely arising as the key issue for a sociologist. Neither should we forget that social hermeneutics ultimately tries to find the deep keys to texts; in other words, it seeks to unveil their inner meaning from an external ideological discourse (Grondin, 2014, pp. 10–1 and 43–107). Consequently, the interpretation of a text additionally implies recognizing its discursive structure, not only at an apparent or superficial level but also in its deep dimension (Mitchell, 2019, p. 23).

2 The Pursuit of Originality, One of the Most Important Characteristics of Creativity

Even though Sociology has traditionally assigned a marginal place to the study of creativity (Joas, 2013, p. 63), which appears as a "slippery and hard-to-define object" (Noya, 2010, p. 19), it has become an emergent research object lately, particularly since Hans Joas published *The Creativity of Action* in 2013. That is perhaps why the sociological approach to its concept can be described as somewhat inaccurate, insofar as it is often addressed on an a priori basis and admitted without reflecting on its true social and cultural meaning. Furthermore, one can frequently check in the Sociology which investigates the issue of creativity that the latter seeks novelty, innovation, and originality, but some imprecision exists there too. By way of example, for E. Durkheim (1987, p. 14), any traditional moral must be altered so that it does not atrophy, and precisely such innovation goes hand in hand with the creation process. According to Hans Joas, creativity attempts to explore ways that are far from routine, whim and everydayness, so that this liberation makes it possible to create and build new actions (Joas, 2013, p. 189 and 254). In the opinion of A. Reckwitz, creativity is now oriented toward the aesthetic and toward the new, the latter being oriented toward the aesthetic (Reckwitz and Sánchez Capdequí 2017, p. 347). The Spanish sociologist C. Sánchez Capdequí (2018, p. 6) in turn believes that the contemporary creative experience almost obsessively pursues novelty and originality; it focuses on the construction of the actual individual and is therefore understood as rupture, difference, and divergence. R. Collins also sees originality and innovation as being inherent to creativity; they represent minority realities and can only bear fruit if they take place under very specific conditions (Rubio-Arostegui et al., 2016, p. 131). From sociobiology, E. O. Wilson (2018, p. 11) claims that creativity implies an innate search for originality.

Because it is not usual for us sociologists to specify what the concept of originality means, this chapter provides an in-depth analysis about its sociogenesis and its inescapable link to creativity. As will be shown, since that originality takes us back to the origin and to the originating, it becomes associated with creation myths, play—the origin of every cultural activity—chaos, and order, as well as with axial civilizations—the germ of our time. Both terms thus prove helpful to understand contemporary society and, more precisely, our hunger for creativity, the goal-centered rationality necessarily complemented by an emotionality that seeks to be intense, and the awareness of reflective society. This becomes representatively and significantly visible in two of the most outstanding creative moments in the history of mankind: in the myths and rites related to the origin of antiquity, which refer to the creation of the cosmos, the world, and living beings; and in contemporary science, which focuses on how life is created and evolves. For their part, both Myth and Science explain the creation process through a complex interaction between vital and sociocultural order and chaos.

3 Etymology of the Concept of Originality: The Link Between the Origin and the Originary Or the Originating

Originalidad [Originality] is defined in Spanish as "*cualidad de original* [the quality of being original]," "*la acción o actitud originales* [the original action or attitude]" (Diccionario de uso del español María Moliner, 2007—DudeMM—) and the "*hecho o rasgo original* [original fact or feature]" (Diccionario del español actual Manuel Seco et al., 1999—DeaMS—). "*Originalidad* [Originality]" refers us back to "*original* [original]" which means "*de origen* [of origin]," to "*la primera de varias cosas que proceden sucesivamente una de otra* [the first of several things which successively come from one another]" (DudeMM), to the "*cosa primera o primitiva, anterior a toda elaboración, cambio o sustitución* [the first or primitive thing, previous to any elaboration, change or substitution]" (DeaMS). Nonetheless, *originalidad* [originality] also has another meaning which contradicts the preceding one, "*lo que no imita a otros* [what does not imitate others]" (DeaMS), what is "*distinto a lo acostumbrado o contrario a lo acostumbrado* [different from the usual or contrary to the usual]," something "*característico* [characteristic]," "*especial* [special]," "*extraordinario* [extraordinary]," "*extravagante* [extravagant]," "*nuevo* [new]," and whatever is associated with words such as "*descubrir* [to discover]," "*explorar* [to explore]," "*innovar* [to innovate]," "*inventar* [to invent]" (DudeMM).

Therefore, originality—the original—leads to the concepts of "origin" and "originary"; that is, both to the first or primitive and to what has been produced in a new, innovative way. The original consequently represents the origin of something and what has arisen innovatively, and also the moment prior to any transformation. In other words, the origin and the originated or originating come together, insofar as the former constitutes the foundation of the latter. However, the intricate connection between these terms does not end here, since the originated in turn has to become the origin of something, a new seed from which a new plant can germinate. Both concepts accordingly merge in such a way that the origin cannot be separated from the originary, and the latter is inseparable from the former, which makes it necessary to understand them using a dialectical method.[1]

This link of creativity with an origin that simultaneously acts as an originator is highlighted by academic literature, which actually characterizes it by an originality that has to do with the origin—and the fascination for genesis—as well as with the novelty of the form, the model, and the new methods (Tatarkiewicz, 1987, p. 292; Weisberg, 1989, p. 15 and 168; Steiner, 2002, p. 25; Nocera, 2009, p. 101) aimed at the future.

Likewise, this same association typically occurs in art. In effect, from the perspective of M. Heidegger (2016, pp. 19–135), art constitutes the origin of the artist and the work, that is, "that from which and for which a thing is what it is and as it is," or, expressed differently, its essence. Therefore, asking about the origin of an artwork means enquiring about the source of its essence or, to put it in another way, about the foundation of truth—founding in the sense of donating, basing, and commencing—which means that the work "sets up a world" and "brings the land here," that it bases the world (a human artistic construction) on the land (the *physis*, nature) and that, conversely, it raises the land by means of the world. Nevertheless, because art inevitably leads to its origin, to its essence, to its foundation, the latter becomes associated with its historical nature, though in the sense of something that belongs to the past but also creates the future, that founds history.

For H-G. Gadamer (1996, pp. 41–2), a disciple of Heidegger, art combines the past, the present, and the future, since it draws its strengths from the art of the past, through a familiarization with tradition, and thus manages to innovate in a daring manner. Similarly, U. Eco (1972, pp. 19–246) considers—following Luigi Pareyson's aesthetics of formativity—that a fully originary intellectual process that is able to invent new laws and rhythms must lie at the basis of every creative act. However, this originality does not appear out of the blue, but arises from the cultural

[1] Human destiny advances dialectically: conceptual dialectics—Hegel: thesis, antithesis, and synthesis, the three stages through which mankind progresses; dialectics of class struggle—Marx; and dialectics of vital reason—Ortega y Gasset (Ortega y Gasset, 2008, p. 149 and 173). For Eugenio Trías, dialectics does not separately study the two concepts at play (subject–object), but the necessary link between them both (Trías, 2019, p. 193); in our case, it would be the connection between the origin and the original. Finally, my interest in dialectics here also relates to its being a methodological option, which above all seeks to identify a reason for change (Eco, 1972, p. 141).

tradition and the physical world within which the artist—trapped between his resistance or his coded, artistically normativized passiveness—receives and freely selects a number of suggestions which are the ones that shape part of the work. Nevertheless, since form represents the final moment in a shaping process as well as the starting point for future interpretations, it explains why the work of art relates to the historical context while simultaneously producing history. Therefore, a peculiar dialectics emerges in the history of art which interconnects the traditional forms with the new ones, in such a way that the formal innovation promoted by art always precedes the revolutions that materialize culturally and socially.

This conceptual ambivalence regarding what originality and its two bordering concepts mean is highly present in the Spanish philosopher Eugenio Trías when he defines, in *La filosofía y su sombra* [Philosophy and its shadow] (2019, pp. 172–9), the human being as a "constituting subject and a constituted object" and as an original being in both circumstances. In other words, a human is not only the one who "owns a previous, original knowledge of his own essence" but also "will be the determination-free subject who will perform an original action able to found 'objects,'" insofar as "human history (is) man's permanent self-building, his continuous objectivation from an original praxis, the ongoing reappropriation of objectivation by the subject."

In short, these two concepts merge in such a manner that the origin cannot be detached from the originary and the latter cannot be separated from the former either. This explains why, in literary, artistic, scientific, religious, or social creation, the alleged originality actually means a return to the origins, without forgetting that, behind its most originary form, creativity announces what is yet to come (Steiner, 2002, p. 94 and 164). For instance, when the artist creates, on the one hand, he establishes some sort of analogy with preexisting items, which means that the originality that he seeks entails a return to the origin, even if its creative act paradoxically seems to erase all previous products. And, on the other hand, this also turns out to be the genesis of a subsequent stream; hence why we should speak about "originating" ideas rather than original ones in this regard. For this reason, the novelty criterion inherent to creativity does not necessarily imply that the response given to a problem, to a situation, or to what has been produced must be objectively unprecedented in the history of mankind, but only new for the creating subject (Llanos, 2018, pp. 2–7).

However, why does that need to return to precedents exist? What role do the latter fulfill in creativity? And how is the origin linked to the originary within that creativity? A possible answer can be found in Science, since—as highlighted by Bohm and Peat (2007, pp. 176–231 and 254 ff.)—creativity is attached to life by virtue of generative order, which becomes essential to understand both life and inanimate matter. Indeed, all that can be accounted for in nature must follow some types of orders, for example, in terms of atoms, molecules, DNA, cells, and other structures. Matter actually arises as a fortuitous combination of molecules which mechanically or in a more or less determined way leads to subsequent developments, which gradually create more complex forms. Therefore, the inanimate context and matter emerge from an overall generative order which comprises every area of experience—science, society, and knowledge. In this sense, they all—and creativity too—are the product of an original vital energy.

At a psychic level, the return to the primeval origin pursues a sort of compensation, as claimed by C. G. Jung. Thus, social relevance lies in the fact that it always serves to educate the spirit of a specific space-time, insofar as the artist's yearning, motivated by a dissatisfaction with the present, makes him unconsciously move backward until he is able to find a propitious original image that can effectively offset both the lacks and the unilaterality which prevail in the spirit of his time. Nonetheless, the source of a symbolic work of art is not in the artist's personal unconscious but rather in the sphere of the unconscious mythology that hosts all the original images, which incidentally constitute a common good of mankind as a whole. Furthermore, this primeval image or "archetype" can be defined as a figure (an essentially mythological one) which, whether it is a *daimon*, a human being or process, has repeatedly appeared throughout history when creative fantasy is exercised; and it results from multiple peculiar experiences of our ancestors. To this must be added that a special emotional intensity can be felt in each repetition, at the very moment when the mythological situation takes place (Jung, 2014, pp. 72–90).

According to Paul Ricoeur (1967, pp. 12–13), "[r]egression offers an alternative route for us to progress and to explore our potential. … Re-immersing ourselves into our archaism undoubtedly provides us with the alternative route through which we can deepen into mankind's archaism, and that double regression in turn possibly represents the path towards a discovery, a research and a prophecy that concerns us."

Finally, J. Ortega y Gasset sees regression as a way to overcome times of cultural crisis.[2] Simplification certainly represents—according to this philosopher—the human method to leave the crisis of culture behind, since the desire for simplicity mechanically incites us to dream about the preceding life, the archaic, initial, or primitive one, the one prior to complication. Hence the interest in returning to the pristine, the nostalgia for primitive life, the eagerness to transform the complicated culture into a simple one (Ortega y Gasset, 2008, p. 143).

In sum, being linked to the creative process, the etymology of the word "originality" takes us back both to the origin and to the originary or the originating and, with them, to genesis and destiny, to the beginning and to the end, to the establishing past and to the institutionalizing future. However, and most importantly, the present in crisis, or felt as insufficient or too complex, is endowed with an original transitory vital energy, with an emotional intensity or—more precisely expressed in sociological terms—with an "effervescence" or a reinvigorated "emotional energy"[3] that comes from the original source. In turn, thanks to its verve, the

[2] Various authors have written on the link between cultural crisis and the emergence of creativity. For example, the scientist J. Monod (Monod, 2007, p. 176) claims that moral anxiety is the source of every creation. Among sociologists, J. Alexander (2017, p. 318) argues that disturbing facts and events bring about new opportunities for innovation and change. According to E. Durkheim (2017, p. 156), "there are historical periods in which, under the influence of some collective disaster, social interactions become much more frequent and active. The outcome is a widespread effervescence, characteristic of revolutionary or creative times." In the opinion of Z. Bauman and Bordoni (2016, p. 13 and 18), the crisis implies a transition period which expresses something positive, optimistic, and creative, and it encourages both a diagnosis of the situation and a call to action. Finally, H. Joas (2013, p. 309) believes that creativity may provide a way out of the cultural crisis.

[3] As is well known, these two concepts—the "effervescence" of revolutionary periods and the "emotional energy" of interaction rituals—are more often used in Sociology. We owe the former to E. Durkheim (2017, p. 156) and the latter to R. Collins (2009, p. 141 ff.). What matters most about both of them, and hence their potential application to creativity, is that they take place at specific moments—during the creative process, in our case—and not on a regular basis.

present can not only build the future but also provide it with a renewed and still unexhausted individual and social complexity. The future thus becomes founded—rooted and legitimized—in the past, whereas the present acts as an intermediary between the other two time frames. Neither must we forget that this contemporary state, by means of the creation process, ends up merging with the remote that underpins it and with the future that it generates, becoming meaningful, intensely experienced, and epistemologically clarified. In this regard, creativity may also prove helpful to remove the confusion which surrounds the present.[4]

4 From the Origin Toward the Originary: The Time Narrative of Creativity

Creativity, a Product and a Producer of Time, of History and Society

It consequently appears that, because it merges all three time frames, creativity arises as a process fully immersed in time and in History, albeit in a dialectical fashion, since they are both makers and products of creativity. Time therefore arguably represents a conscious form of life and, conversely, life develops the temporal dimension (Simmel, 2017, p. 219). And of course, time also defines the human being; in the words of F. Schiller (1990, pp. 195–7), "man … is an individual who finds himself in a specific state. But every state, every given existence has its origin in time, and thus, man as a phenomenon must have a beginning. … Without time, or expressed differently, without the possibility to evolve, he would never be a specific being." It consequently comes as no surprise to check that the human being implies creating time, which simultaneously entails alterity and destruction (Castoriadis, 2017, pp. 39–42) and, likewise, that civilization sets up a creative and destructive process, which builds further civilization and, at the same time, causes a decivilizing

[4]On the current state of confusion, see the book *En los límites de la confusión: miedos, riesgos y urgencias de la Sociedad de la Información* [In the limits of confusion: fears, risks, and urgencies of Information Society"], by Celso Sánchez Capdequí, 2010.

effect (Elias, 2000, p. 16). Alongside this, time becomes a basic component of the social sphere, one of its constituent parts, the one dynamizing it, the engine that continuously sets it in motion (Balandier, 2014, p. 62). This social time always has an imaginary nature, though, and is imbued with signification (Castoriadis, 2017, p. 32); hence its connatural creativity, since we should not forget that the imaginary is the actual instrument which serves to criticize reality (Ricoeur, 1988, p. 104) and that it may represent a second life, is created and superimposed on the real one (Lizcano, 2003, p. 10).

Lastly, time has a historical dimension, History—the same as time—standing out as a key concept of society, and particularly of sociability. In this sense, it is worth highlighting—as Novalis pointed out—that History, like creative originality, comprises the past and the future, the memory and the hope (Koselleck, 2016, p. 82 and 126), and also that time precisely inhabits the present because it is in our mind, as a memory and as an anticipation (Rovelli, 2018, p. 135). This means that, as far as human beings are concerned, History clearly represents the past, the sphere where the preterit is locked (Jaspers, 2017, p. 53), without this being a hindrance for it to own a contemporary ontological dimension: today "is" (Bellah, 2017, p. 12). The past is certainly the present, since the latter is made using the matter of the former, in such a way that we constitute its summary (Ortega y Gasset, 2008, p. 55). However, at the same time, the past represents an awareness of the future (Jaspers, 2017, p. 53). In fact, we reminisce about it precisely because we expect the future taking the past into account (Ortega y Gasset, 2008, p. 120). Within this context, History has a transitive nature, especially because it does not express what is typical of it or what survives, but the unique creations, irruptions, and realizations and, ultimately, what cannot be repeated or replaced. Therefore, the present irrevocably needs to be perceived both from the future and from the past (Jaspers, 2017, p. 210 and 356–367), to such an extent that if the past does not exist, neither do the other two (Bellah, 2017, p. 12).

The present sphere thus has blurred, open, and widened boundaries because of what precedes it and what follows it. Instead, what has already happened does show a fixed, immutable, and (relatively) unequivocal figure; and the future is characterized by problematic, plural, undetermined, random, and insecure limits which make it impossible to establish a stable ground. For that reason, in the future, human beings cannot find their bearings without looking back at the past (Ortega y Gasset, 2008, p. 120). Nor can they live in the present, making sense of it, unless the present goes hand in hand with memory and hope.

In any case, at least three consequences result from the temporal and historical dialectics that I have outlined here:

(1) It is impossible to understand the present from the past unless one has entered the latter willing to understand the present (Mannheim, 2018, p. 170), or to put it in another way, the idea is to apprehend the present through the past and, correlatively, to comprehend the past via the present (Ricoeur, 2010, p. 223). Hence our interest as sociologists in the historical method, insofar as it helps explain the current situation (Beltrán, 1985, p. 9). It follows from the above that society is the result of a far-reaching social development, the interpretation of which must be integrated into the global evolutionary process (Elías, 2000, pp. 14–83). Certainly, the human group stems from a long bio-human evolution (Veblen, 2008, p. 20), from the shaping of Nature and History, the first materializing in genetic inheritance and the second in tradition[5] (Jaspers, 2017, p. 346),

---

[5] According to G. Balandier (2014, p. 37 and 87), three main traditional modalities exist: (1) fundamental traditionalism, which keeps the most deeply rooted values, models, and social and cultural practices, and places itself at the service of permanence; (2) formal traditionalism, which uses conserved forms with a modified content and preserves the continuity of appearances, albeit with innovative aims; furthermore, every movement that occurs is united to the relationship with the past, which remains preserved; and (3) pseudo-traditionalism, which can be described as a reformed tradition developed in those stages during which change accelerates and major upheavals take place; it likewise grants meaning to the new and unexpected, as well as to change, which it manages to tame by imposing upon it a familiar or comforting appearance, and expresses an order born from disorder. In this respect, the work of tradition cannot be separated from that carried out by History, since the former constitutes a reserve of symbols, images, and means which permits to mitigate modernity.

On another note, this same scholar suggests that, in accordance with the Western conception, tradition has two representations: one that is passive and has as its main aim to preserve memory; and the other, active, which makes it possible to build what has already existed. These two representations use the word, the symbol, and the rite, which own this double-faced nature—active and passive. For all these reasons, tradition becomes embedded in a history where the past continues into the present and whose truth comes from world order since the origin.

understood not as a mere conservation, but as a transmission (Gadamer, 1996, p. 116).

Therefore, instead of fighting one another in a polarized manner, the biological and social processes, the natural and social spheres, and the social and individual forms must become intertwined so that they can become more effective, without forgetting that it is precisely the unforeseen and hazardous technical creativity of the evolutionary process that generates such an interdependence (Elías, 2000, pp. 14–83).

(2) "Nothing is lost forever," since the previous stages are preserved, reorganizing themselves under new conditions. According to Science, every life constitutes a development of the first substantial process, that of the prokaryote cell, where stability and change strengthen each other instead of becoming antagonists; after all, if mutations are random, they have to be affected or rejected by the preserved substantial processes. It is thus these processes that cause variation, that is, the "facilitated variation," in such a way that an equilibrium, or a dialectic, arises from the most remote past between the maintained structures and the innovative variations (Bellah, 2017, p. 22 and 49, 104 and 135). This dialectic additionally triggers a constant rejuvenation of society; not in vain, the key present-day transformations take place through the dissolution of its previous form and the building of a new one, without the latter ever looking like a radically different construction, as it brings together the old and the new within the transformation process. This happens because, while building the new, we rediscover certain elements in the traditional context that open up previously unappreciated and unexplored possibilities, and this takes place even in reforming or revolutionary movements in which new virtues become linked to old ones (Mannheim, 1993, p. 31 and 214 and 5).

(3) The historical dialectic to which I am referring derives from the polarizations that are present in a given period. These polarizations constantly orient themselves with respect to one another and are mutually related by means of a dynamic tension which shapes the dynamic–antinomic unity of that specific stage (Mannheim, 1993, p. 234). Among such polarizations stands out the one that pits

continuity against discontinuity—that which generates the present and the future (Bauman & Leoncini, 2018, p. 106)—and those which, following Bourdieu's theory (Rubio-Arostegui et al., 2016, p. 129), produce creative innovations, tradition, and avant-garde, the classical and the modern, together with, as will be seen later on, order and chaos.

Summing up, time is connatural to life, to the human being, to civilization, to History, to society and to creativity itself. If society, through its prolonged duration, weaves a narrative plot that inextricably joins it to time, so does creativity without a doubt. After all, creativity is immersed in temporality, in History, and in society, and it is both the product and the producer of all three; what is more, creativity arguably represents an essential factor in the interdependence existing between them. This is so because it—creatively—uses a dialectical temporal process to solve the tension which characterizes the relationship between tradition and avant-garde, continuity and discontinuity, and their social agents, both the old and the modern ones. And, ultimately, because by acting in this way, it tries to embrace memory and hope, to critically recall collective memory, and to build new utopias at the same time. Consequently, creativity intensifies, revitalizes, and rejuvenates the sense of a present located between the origin and the originary.

Bibliography

Alexander, J. (2017). Trauma cultural e identidad colectiva. In C. Sánchez Capdequí (Ed.), *La creatividad social: narrativas de un concepto actual* (pp. 315–347). CIS.

Balandier, G. (2014). *El desorden. La teoría del caos y las ciencias sociales*. GEDISA.

Bauman, Z., & Bordoni, C. (2016). *Estado de crisis*. Paidós.

Bauman, Z., & Leoncini, T. (2018). *Generación líquida. Transformaciones en la era 3.0*. Paidós.

Bellah, R. N. (2017). *La religión en la evolución humana. Del Paleolítico a la era axial*. CIS.

Beltrán Villalva, M. (1985). Cinco vías de acceso a la Realidad Social. *REIS, 29*, 7–41.
Beltrán Villalva, M. (2012). Sobre hermenéutica: de la Filosofía a la sociología empírica. *RES. Revista Española de Sociología, 17*, 9–26.
Beltrán Villalva, M. (2016). *Dramaturgia y hermenéutica: para entender la realidad social*. CIS-Centro de Investigaciones Sociológicas.
Bohm, D., & Peat, F. D. (2007). *Ciencia, orden y creatividad. Las raíces creativas de la ciencia y la vida*. Kairós.
Castoriadis, C. (2017). Tiempo y creación. In C. Sánchez Capdequí (Ed.), *La creatividad social: narrativas de un concepto actual* (pp. 21–47). CIS.
Collins, R. (2009). *Cadenas de rituales de interacción*. Anthropos.
Domínguez, J. (1997). Estética hermenéutica y hermenéutica de la imagen. La articulación de imagen y lenguaje. In H. G. Gadamer & G. Boehm (Eds.), *Estudios de Filosofía* (Vol. 15–16, pp. 91–111).
Durkheim, É. (1987). *Historia de la educación y de las doctrinas pedagógicas*. La Piqueta.
Durkheim, E. (2017). La efervescencia religiosa. In C. Sánchez Capdequí (Ed.), *La creatividad social: narrativas de un concepto actual* (pp. 151–169). CIS.
Eco, U. (1972). *La definición del arte*. Ediciones Martínez Roca.
Elias, N. (2000). *Teoría del símbolo. Un ensayo de antropología cultural*. Península.
Fernández Galán, C., & García Ramírez, J. (2012). Lectura de la imagen: ¿semiótica o hermenéutica?. *Imaginario Visual Investigación • Arte • Cultura*. Año 2, No. 4. Noviembre–Abril 2013, pp. 48–55.
Gadamer, H.-G. (1996). *La actualidad de lo bello*. Paidós.
Gadamer, H.-G. (1997). *Mito y razón*. Grupo Planeta (GBS).
González García, J. M. (1992). *Las huellas de Fausto. La herencia de Goethe en la sociología de Max Weber*. Tecnos.
González García, J. M. (1998). Sociología e iconología. *Madrid: REIS, 84*, 23–43.
Grondin, J. (2014). *A la escucha del sentido: conversaciones con Marc-Antoine Vallée*. Herder.
Heidegger, M. (2016). *El origen de la obra de arte*. La Oficina.
Jaspers, K. (2017). *Origen y meta de la historia*. Acantilado.
Joas, H. (2013). *La creatividad de la acción*. CIS.
Jung, C. G. (2014). *Sobre el fenómeno del espíritu en el arte y en la ciencia*. Trotta.
Koselleck, R. (2016). *historia/Historia*. Trotta.
Lince, R. (2009). *Hermenéutica, arte y ciencia de la interpretación*. UNAM.

Lizcano, E. (2003). Imaginario colectivo y análisis metafórico. transcripción de la conferencia inaugural del *Primer Congreso Internacional de Estudios sobre Imaginario y Horizontes Culturales* que se celebró en la Universidad Autónoma del Estado de Morelos, Cuernavaca, México, del 6 al 9 de mayo. Retrieved October 7, 2019, from http://www.unavarra.es/puresoc/pdfs/c_salaconfe/SC-Lizcano-2.pdf

Llanos Melussa, E. (2018). La creatividad, ¿qué es y cómo podemos estimularla? *Creces*, 6–14.

Mannheim, K. (1993). El problema de las generaciones. *REIS, 62*, 193–242.

Mannheim, K. (2018). *El hombre y la sociedad en la época de crisis*. CIS.

Mitchell, W. J. T. (2019). *La ciencia de la imagen. Iconología, Cultura Visual y estética de los medios*. Akal.

Moliner, M. (2007). *Diccionario de uso del español Maria Moliner*. Gredos.

Monod, J. (2007). *El azar y la necesidad*. Tusquets.

Muñoz, B. (2001). Reflexiones sobre la sociología de la cultura y de la música en la obra de Max Weber: un análisis crítico. *Sociedad y Utopía. Revista de Ciencias Sociales, 18*, 23–37.

Nocera, P. (2009). Los usos del concepto de efervescencia y la dinámica de las representaciones colectivas en la sociología durkheimiana. *REIS, 127*, 93–119.

Noya, J. (2010). Sociología de la creatividad. In J. Noya, F. Del Val, & M. Pérez Colman (Comps), *Musyca. Música, sociedad y creatividad artística*. Biblioteca Nueva.

Ortega y Gasset, J. (2008). *En torno a Galileo*. Revista de Occidente-Alianza.

Reckwitz, A., & Sánchez Capdequí, C. (2017). La estetización y el dispositivo de la creatividad: el régimen social de la novedad estética. In *La creatividad social: narrativas de un concepto actual* (pp. 347–374). Centro de Investigaciones Sociológicas (CIS).

Ricoeur, P. (1967). *The symbolism of evil*. Harper and Row.

Ricoeur, P. (1988). *Hermenéutica y Acción. De la Hermenéutica del Texto a la Hermenéutica de la Acción*. Prometeo Libros-UCA.

Ricoeur, P. (2010). *La memoria, la historia, el olvido*. Trotta.

Rovelli, C. (2018). *El orden del tiempo*. Anagrama.

Rubio Arostegui, J. A., Pecourt, J., & Rius Ulldemolins, J. (2016). Usos y abusos de la creatividad. Sociología de los procesos creativos, transiciones a lo digital y políticas creativas. *Debats, 130*(2), 125–145.

Sánchez Capdequí, C. (2010). *En los límites de la con-fusión. Miedos, riesgos y urgencias de la Sociedad de la Información*. Libros de la Catarata.

Sánchez Capdequí, C. (2018). Las sociologías post 68 y la creatividad social. *Arbor, 194*(787), a435.
Schiller, F. (1990). *Kallias; Cartas sobre la educación estética del hombre.* Anthropos.
Simmel, G. (2017). La trascendencia de la vida. In C. Sánchez Capdequí (Ed.), *La creatividad social: narrativas de un concepto actual* (pp. 213–229). CIS.
Steiner, G. (2002). *Gramáticas de la creación.* Siruela.
Tatarkiewicz, W. (1987). *Historia de la Estética. I. La estética antigua.* Akal.
Trías, E. (2019). *La filosofía y su sombra.* Galaxia Gutenberg.
Veblen, T. (2008). *Teoría de la clase ociosa.* Alianza.
Weber, M. (1971). *Sobre la teoría de las ciencias sociales.* Península.
Weber, M. (2006). *Conceptos sociológicos fundamentales.* Alianza.
Weisberg, R. W. (1989). *Creatividad. El genio y otros mitos.* Editorial Labor.
Wilson, E. O. (2018). *Los orígenes de la creatividad humana.* Crítica.

Part II

Creative Categories of the Origin

Chapter 2, which begins Part II dedicated to the creative categories of the origin, tries to provide a specific description of the term from its genealogy regarding the sociocultural–historical context where it arises, from a comprehensive Weberian sociology and the social heuristic method. More precisely, it deals with the idea conveyed by Western myths: the narrative about the identification of the instituting generatrix forces or the procreative divinities that lie behind the birth of the cosmos, of the world, of society, of the earth, of gods, of humans, of animals, and of plants. Thus, from an interpretative examination of the myths about Mother Goddess and Biblical Genesis, as well as of Greek creation myths, an attempt is made to draw a conceptual map that delimits the most defining features of creativity (1). The ultimate goal is to check whether such characters have survived to the present day (2).

As seen in the first chapter, the originality of creativity refers, etymologically, to the neighboring concepts of "origin" and "originary." With them, creativity generates a temporal narrative that inseparably interrelates the past, the present, and the future. This temporal narrative is observed—as shown in Chap. 3—in a representative and significant way, in two of the most intense creative moments in the history of humanity: in ancient myths and origin rites, which refer to the creation of the cosmos; and in contemporary science, which focuses on how life is created and evolves. In both creative moments there is the dialectic between chaos and order, linked to the memory of the past and the projection into the future. But this dialectic between chaos and order has changed from the first to the second modernity, because now it has become more complex and convoluted, without being able to clearly distinguish the borders between what is chaos and what is sociocultural order.

Chapter 4 tries to specify, first of all, to what extent the axial civilizations were produced around 500 BC. To do this, I will investigate the etymology of the concept of originality and its connection with the past, the present, and the future. As will be seen, this is what takes place in the Axial Age, one of the most intense creative moments in human history, in which there is a close link between play and the creation of axial utopias. To this end, I will develop two conceptual maps that delimit the characteristic aspects of both the game and the utopia. Thus, it will be verified that the Axial Stage is, indeed, original and that it organizes a highly creative temporal narrative that renews the tradition of the past while building our future.

Finally, Chap. 5 ends the part dedicated to the creative categories of origin. It seeks to reveal that creativity is the result of the conflict between generations and, more specifically, between childhood and youth, characterized by innovation, and old age, defined by its conservative nature. This generational clash is observed, in a particular way, in the art and science of the first modernity, since they constitute a deeply creative childhood or youth stage. On the contrary, the artistic and scientific creation of the second modernity has entered a generalized physical, cultural, and social decline. Thus, it is explained that youth have lost prominence in contemporary society and, in short, that it has aged. This has been the result, among other causes, of the consequences of the world wars, demographic aging, the decline of the arts and sciences, and the attenuation of the generational conflict between youth and old age.

2

Mythical Bases for a Sociological Definition of the Concept of Creativity

1 Introduction

Creativity has a set of foundations—anthropological ones, as they would be referred to by H-G. Gadamer (1996: 32)—upon which its legitimacy rests. Following the etymology of the word "originality," characteristic of creativity, it is my understanding that these two foundations, or categories as I prefer to call them, would be distributed in two broad sections: those related to the origin; and the ones which are linked to the originary or the originating. As for the former, an outstanding role corresponds to those which have to do with the myths of creation, chaos–order, childhood and youth, play, and axial civilizations.

Since modernity arises as a process whose dynamics draws upon religious roots (Knöbl, 2017: 285), it seems appropriate to look for the bases of the concept of creativity in those roots. For this reason, I will dedicate this chapter to analyze the Western myths which narrate what instituting driving forces or procreating divinities stand behind the birth of the cosmos, the world, society, the earth, gods, human beings, animals, and plants. Thus, from an interpretive examination of the Greek myths of Mother Goddess, of Biblical Genesis, and of creation, a concept map will

J. A. Roche Cárcel, *Creativity and Time: A Sociological Exploration*, Palgrave Studies in Creativity and Culture, https://doi.org/10.1007/978-3-030-84838-5_2

emerge for the purpose of establishing the key features of creativity; the ultimate goal is to check whether these features of the origin become originators and, therefore, whether they survive at present or not.

To the theoretical and methodological foundations announced at the beginning of this book must be added the references within the framework of M. Weber's civilization stages and the religious-order stages described by Freud. Nevertheless, given the limited importance that these two scholars assigned both to creativity itself and to the role played by women in it as well as in civilization, attention will likewise be paid to such influential researchers as M. Gimbutas, J. Campbell, D. Boorstin, G. Steiner, and others and, especially, to the postulates of other sociologists who dealt with the evolution of civilization and with the role played therein by religion and family structure.

For Weber, the History of Mankind has developed in three main stages: that of tribal societies and neighborhood ethics; the prophetic or salvation era of fraternity ethics; and the modern individualistic, capitalist, or rationalist society. Three social agents act successively in the three aforementioned stages—the wizard or shaman; the prophet, the savior, the ascetic, and the mystic; and the charismatic man—within a social structure that is rooted in kinship and neighborhood (Bellah, 2017: 279–300). Today's higher knowledge, in its most developed form, is a product that derives from the priestly office and the idle life, whereas modern science results from the industrial process. Consequently, the sages, scholars, inventors, and speculators, a majority of whom have carried out their most significant works outside the protection of schools, the same as the mental habits imposed by modern industrial life, have been coherently expressed through a body of theoretical scientific knowledge (Veblen, 2008: 377).

The same as Weber, Freud (1985: 37–198) believes—in *Totem and Taboo*—that family organization, together with the religious factor, constitutes the key to historical development. Thus, the paternal horde is replaced by the fraternal clan, in parallel to the substitution of the Son's religion for that of the Father. According to both Weber and Freud, these evolutionary perspectives stem from the religious-family structure and they both obviate the role of women, even though Freud (1985: 153) recognizes that totemism is a creation of the "feminine spirit."

The first acknowledgment of a matriarchal order in the Near East and Europe dates back to 1861, in J. Bachofen's *Mother Right* (Engels, 2013: 49 ff.). However, it was M. Gimbutas, with her compilation, classification, and descriptive interpretation of nearly 2000 symbolic artifacts coming from European Neolithic sites (7000 B.C.–3200 B.C.), who provided us with the first great systematization of the "Language of the Goddess." This implies a huge historical breakthrough, since it has revealed a philosophy of human existence opposed to the highly ideologized, hierarchized, and manipulated systems which have prevailed on an absolute basis in the Western World during the successive historical periods (Campbell, 2015: 387–389).

It thus seems logical—and ethical—to claim that creativity is also feminine and, accordingly, to divide the myths about the origin into three distinct areas: that of Mother Goddess, corresponding to a hypothetical matriarchal society; that of Father God, related to a patriarchal society; and that of Son God, linked to a fratriarchal and democratic era. This tripartite periodization is mentioned by A. Ortiz-Osés (1993) and a variety of sociologists, including J. Beriain (1999: 70–86), J. A. Bergua (2015: 133 ff.), and M. Maffesoli (2004: 29, 37, 89), which provides evidence of its obvious sociological relevance. Furthermore—as seen later—it proves very useful to find the origins of Western creativity or, expressed differently, the creation of our civilization.

Summing up, taking the aforesaid theoretical and methodological bases as our reference, this article is structured around three fundamental sections—respectively dedicated to the genealogy of creativity offered in the myths of Mother Goddess, Father God, and Son God—preceded by an introduction and culminating with a conclusion which describes the way in which the mythical bases of creativity remain present in sociological thinking at present.

2 Mythical Genealogy of the Concept of Creativity

Creativity in the Religion of Mother Goddess, Creatrix of the Universe

The Goddess of Life and Death: Generatrix and Nourishing Mother and Womb for the Return of the Dead

The Great Goddess, who stands out as the essential divine figure in the world's first mythological cosmovision (Campbell, 2015: 14), originates in the Paleolithic but deploys in the Neolithic. The Paleolithic witnessed the oldest female artistic representations—circular or pear-shaped vulvas, a symbol of sex (Delporte, 1982: 8 and 247; Gamble, 2001: 319)—which express the conception that hunting societies have about women and divinity, and also constitute antecedents for the later myths which preserve the old conceptions of the Goddess—first orally and then through writing. A special mention must also be made of the Aurignacian–Gravettian Venuses, small sculptures found in a broad area—France, Italy, Central Europe, Russia, and Siberia (Delporte, 1982: 29, 218 and 311). In general, this early naked women-goddesses simultaneously represented the origin itself, the matrix of future mankind as a whole and the omnipresent sex in a wild state (Lessing & Sollers, 1994: 33), without forgetting that they likewise embody an obese and adipose woman-goddess with an idea of abundant beauty, right when survival was a question of food and fertility (Lucie-Smith, 1994: 11–12). These Venuses have been located outdoors, where families lived (Campbell, 2015: 19–20), but also inside the deep caves, away from the inhabited everyday space (Delporte, 1982: 210 and 293–299). It is not by chance that these caves symbolize the vulva, the matrix of Mother Earth and the entry or the return to the maternal uterus (Gimbutas, 1991: 233 ff.; Gimbutas, 1997: 47; Husain, 2001: 54 and 163; Smith, 2003: 197 and 230–231; and Campbell, 2015: 45).

The Goddess of the first farmers develops at a time when the main concerns are no longer animal hunting and slaughter but sowing and harvesting. The oldest images of the Great Goddess of the farming cultures come from the Near East and Europe, and date back to the period comprised between 7000 and 5000 B.C. (Campbell, 2015: 24–25). Nonetheless, the relationships between humans and nature had already dramatically changed around the year 10000 B.C., after they domesticated nature and subsequently dominated it, which in turn brought about a cultural, ideological, and mental transformation of the human beings themselves (Guilaine & Zammit, 2002: 101).

Neolithic religion granted a dominant role to matriarchal symbols, which express the analogy between human and plant fertility and their religious rites (Eliade, 1989: 127), as well as the close and mystical bond between the human being and the common (Cirlot, 1988: 30 ff) deep rhythms (Ries, 1997: 13), on a permanent evolution, of a nature deified by its cyclical, renewing, and generative potential (Eisenstadt, 1986: 1–25; Gimbutas, 1997: 58). Along with this, the existence of a Pregnant Goddess of Vegetation expresses the idea of Mother Nature with an impregnated womb, recognizing that the seed is the cause of germination and the reason for the swollen belly of a woman to be assimilated with a sown field. It additionally stresses that the forces of the earth concentrate in mounds, hills, rocks, and trees, and that they associate their matrix with the curvilinear outlines of the land and with the cave (Gimbutas, 1991: 233 ff.; 1997: 47), the first city, that of the dead (Mumford, 1979: 11). Therefore, Mother Goddess embodies life and death and the afterlife that souls emigrate to; it consequently personifies the psychology of mortality and resurrection in which the agrarian mindset is rooted (Bru, 1990: 23–25).

In sum, Mother Goddess has two basic functions: she grants life and receives you in death. She is the goddess who gives birth, who takes the seed and turns it into life, but—as the farming people knew—the seed grows, develops, matures, dies, and finally returns to the earth from which it came. A third and quite important one for our topic should be added to these two basic functions, though: Mother Goddess inspires poetic realization and the wishes to be not only animals (Campbell, 2015: 73–87). In short, the Mother Goddess of matriarchal society is the Creatrix of the Universe and the poetic muse of humans.

Features of Mother Goddess's Creativity

But what are the outstanding characteristics of creativity according to Mother Goddess's mythology?

1. The identity between Creatrix and Creation expresses unity, insofar as creativity is a generating flow that goes from the mother to her children, or from the creatrix to her creations, and vice versa. No breaking of the umbilical cord consequently takes place, since this creativity owns the will to unite what is separated, to erotically merge the opposed strengths born on the Earth. Hence, neither do divisions, hierarchies, genders, competitiveness, or conflicts exist in Mother Goddess's creativity; nor does diversity, even though—it must be remembered—Mother Goddess diversely manifests herself always remaining unique, as her body represents the cosmos that combines various states of being (Husain, 2001: 6–45). Consequently, Mother Goddess entails the illogical—dialectical—coexistence of opposite ends, insofar as she is at the same time the Mother of Mankind, of animals and plants, earth and sky, death and renaissance, light and darkness, and peace and war (Dunn, 2008: 14).
2. This merging and erotic creativity is associated with fertility and fecundity, related to the woman and the source of life (Flynn, 2002: 25), to vital strength. Furthermore, since she is solidly seated in a throne—the Goddess of Çatal Hoyuk, Cybele, Isis, and the older Hecate—the same as a mother, she protects and gives security.
3. This explains why Mother Goddess is Nature and her creativity not only stems from it but also returns to it; and hence the raw material used in her generatrix action is physical, carnal, and corporeal. That is how it should be, Mother Goddess sculpts—as if she were a sculptress—a body which is "earth" and not "smoke" and does so through an organic earth teeming with life and enriched by nourishing humus formed by the decomposed bodies of the dead.
4. The itinerary from life to death which could be described as circular is associated with a creativity that owns a human temporality parallel to nature, since it has a cyclical nature, marked by continuity and governed by the seasons and the passing of days (Duque, 2000: 44). It

is actually characterized by the phases of the moon (new, full, and waning), which match the life cycles of the matriarch, of the woman (the maiden, the nymph or nubile woman, and the elderly woman) with the sun seasons (spring, summer, and winter). The time of maternal creation is equally that of the "return to the seed" (Gallego, 2003: 379), that of remembrance (Jünger, 1998: 68 ff), and that of the periodical return to the past, the one which inspires the feeling of nostalgia for a repeated return to the mythical time of the origins (Eliade, 1952: 11).

Creativity in the Religion of Father God and of Patriarchal Society

The Mother's Curse and the Establishment of Father Creator

While the civilizations of the Great Goddess in the Near East and Europe essentially obtain their food from farming, nomadic tribes that breed cattle travel across the vast neighboring lands of the Syrian desert, the European and West Asian plains: in the south, Semitic herders of sheep and goats; and in the north, a wide variety of Indo- European peoples. These warrior tribes are not made up of peace-loving and persevering peasants but of nomadic horsemen, whose tutelary gods—Marduk, Ashsur, and Yahve among Semites—cause thunder and instill fear. In successive waves, from the fourth millennium, these Indo-Europeans move to Mother Goddess's areas, where they marry their gods to the local female deities, whereas Semites lay waste to everything that lies in their path from the Syrian–Arab desert to Canaan and Mesopotamia. Indeed, the latter show a much more radical behavior when it comes to the removal of autochthonous goddesses. This actually results in a complete victory of the masculine principle over the feminine one, in such a way that Mother Goddess is reduced to her lowest level: she represents the chaotic, abysmal, and informal cosmic water through the surface of which the human personification of God's shaping spirit moves (Campbell, 2015: 30–153).

As opposed to the static and peaceful culture of Mother Goddess, that of Semites is unstable and transhumant. Furthermore, their gods are masculine and their cosmos has to do with the celestial sphere and not with the interior of the earth, which is why salvation does not lie in its entrails, but in the ascent to heaven, the symbol of the Supreme Father. Added to this, if agrarian religions presented the nudity of the goddess and the sexual act as sacred, religion now moves away from frenzied and orgiastic sensuality, and the sexual act becomes sinful (Bru, 1990: 26).

Moreover, Jews' history is marked by almost 2000 years of exile—as attested by the Bible, which provides an accurate narration of the exodus from Egypt. The diaspora starts in 70 A.D., which means that the banishment lasted over 1880 years, since it officially came to an end with the foundation of the State of Israel, in 1948 (Hayoun, 1995: 201–202). Most interestingly, the Bible is solidly rooted in an earthly story, insofar as it constitutes a drama where the people of Israel play the starring role and dictate the world's fate (Finkelstein & Silberman, 2005: 1–16).

Features of Father God's Creativity

1. Yahve, "the one who makes existence possible," "the Creator" (Boorstin, 2008: 44 ff.), unlike Mother Goddess, does not start from the visible and sensitive body and, therefore, his creation is not erotic, but rhetorical, since it comes from the mind and, more specifically, from the word, the most important maker of the universe (Steiner, 2002: 46; Boorstin, 2008: 17): "God said, Let there be light, and there was light" (Genesis 1, 3–5). This explains why God—and creativity—acquire a nonmaterial, abstract, and exclusively spiritual dimension.
2. Nonetheless, God firstly and paradoxically generates the primary material elements: "In the beginning, Elohim created the skies and the earth" (Genesis 1,1). The state in which the latter found itself before the first day's work consisted in a shapeless mass, with no order whatsoever and without any vegetation, covered by waters and immersed in darkness, on which the Spirit of God alighted (Monforte, s.a). Thus, prior to the organization of the world, only confusion and

disorder reigned, *tehom*, a chaotic mass or primordial matter which—the same as in Mesopotamia (Monforte, s.a)—represents a distant echo of the old Mother Goddess, through which the Creator floats. For that reason, an ever-present wild mystery continues to concur in the original creation, at the deepest, in the genesis of every work of art, of any philosophical intuition or of all sorts of creativity, inasmuch as it distils some "other" inhuman, some inhuman, "otherness" (Steiner, 2002: 57–58).

3. Anyhow, God creates *ex nihilo*, since there is no pre-existing materiality, which explains why creation does not equal mimesis—which will happen with the Greeks, as shown later.
4. The Creator is the Whole, the complete and meaningful being, a Unique, Absolute, Infinite, and Eternal being; and, consequently, his Creation constitutes a unity conceived with utmost care where each detail assumes a function, a characteristic relationship with the whole (Frye, 1988: 138). However, if He embodies the homogeneous Whole and the Unity, his Creation conversely represents the part and the difference and, accordingly, his creativity could be based on nothing but differentiation and hierarchization (Frye, 1996: 207); this actually explains why angels emerge first, followed by men, and, finally, by Nature. Furthermore, the contrast between the created world and the initial chaos predetermines a dual, polarized creativity which opposes day and night, heaven and earth, flesh and spirit, world and dream, as well as perfection and imperfection.
5. Polarization, reflection, logic, and linearity precisely shape the masculine methods *par excellence* (Dunn, 2008: 13) and, indeed, when the masculine principle of Father God comes into play, division, separation, and differentiation take place. Instead, the presence of the Mother Goddess feminine principle—as seen earlier—immediately gives rise to union or fusion.
6. The story of Eden attempts to transfer the social power of the pre-Biblical earthly goddess to a symbolically masculine Father God associated with the skies. The latter is therefore conceived as paternal, in contrast to the maternal Nature, thus transforming the mother into the relative from whom the human being must detach himself in order to be born—physically and spiritually. When she becomes

assimilated with Mother Nature, we find a worship of Goddess Earth as the exponent of a human society in a nascent state and "locked" within a natural cycle of life, death, and renaissance. However, the moment when Adam falls, he is expelled from Paradise, from Nature, which has lost its divine aura (Frye, 1988: 93), and is mythologically defamed—the curse of the earth, *adamah*, grammatically feminine in Hebrew (Cencillo, 1970: 65, 123 and 230); the curse of the snake, the animal of Mother Goddess (Billinghurst, 2007: 22); and the condemnation of Eve, the first woman, the guardian of chaos like the ancient Mother Goddess (Frye, 1996: 189, 254 and 256; Campbell, 2015: 34). Therefore, the whole Old Testament history has an antinatural tone, insofar as it consists in the confrontation between Yahve and the cults of Nature, which implies the most extreme attempt to abolish it.

7. What is more, instead of Nature, the Bible emphasizes the social sphere or social laws—all that has to do with what men do to stay together. In this sense, the social dimension of Father God's creativity becomes manifest in the political, "nationalist" aspect (Finkelstein & Silberman, 2005: IX ff.) of Father God's mythology, and not only because it reflects the patriarchal social family structure of Jewish society but also because it constitutes the mythological defense, the idealization of a people forced to banishment from their territory. Hence—as the Bible says—"There is no God in all the earth, except in Israel" (2 Kings 5:15), a dogma which sharply contrasts with the meaning of the Goddess, who can be found in every territory and who is any place (Campbell, 2015: 34), that is, a universal, nonethnic goddess.
8. Along with that social dimension of creativity, the Bible also states its individual power, symbolically masculine as opposed to the communal, which is a feminine creation (Frye, 1996: 266). For this reason, the God of *Genesis* is alive, creative, and personal (Hayoun, 1995: 205); and his singularity precisely lies in his power as a maker—as seen earlier—which means that the peculiarity of human beings, what builds them as such, is therefore their ability to imitate God and his way of exercising the power of creation (Boorstin, 2008: 71).

9. Divine creativity was essentially free, since it gave rise to the world by means of a purely volitive act (Hayoun, 1995: 201). In fact, the Creator decided to create, and he might have chosen to live in the uncreated infinitude of the void of the neighboring chaos, the only companion during his existence. Thus, biblical creativity always has an optional, contingent, and free nature (Steiner, 2002: 134–136). Deeply linked to this divine creative freedom lies its most outstanding creature, the human being, who has actually been endowed with free will by God.
10. Temporality is an invention of Judaism inherited by the Old Testament which constitutes "the religion of time," a true "builder of time" (Beriain, 2000: 36; 2008: 43). Thus, in the myth of *Genesis*, the world is created in seven days with a huge effort, which leaves God so exhausted that he needs a day's rest. After all, Creation represents a historical event, the prolog to the rest of History (Boorstin, 2008: 24 and 71), the absolute beginning and end of time and space (Frye, 1988: 96). It is in *Genesis* that two births created by God—that of cosmos and that of time—are narrated, and where an explanation is given about the process through which we go from the present to the future being; that life, like temporality, results from nothingness, and that history moves. It follows from this that the world is neither eternal nor immutable; on the contrary, the world is immersed in time and has been created in such a way that it becomes a historical product, a means for human action, that it leaves behind the idle and inactive universe of Paradise (Cioran, 1986: VI), and that it is determined by uncertainty, insecurity, and freedom (Beriain, 2000: 48; 2008: 44–46).

Creation in the Religion of Son God or of Democratic Fratriarchal Society

Greece: From the Matriarchal Substratum to the Patriarchal One

The history of Greece develops in three stages from 800 or 750 to 500 B.C.—the archaic period; the fifth and fourth centuries—the classical period; and from Alexander the Great to the Roman conquest—the Hellenistic period (Finley, 1975: 29). This history is influenced by the unstable conception of the world and the human being (Castoriadis, 1998: 70 and 115 ff.; Rodríguez Adrados, 1983: 386–422; Mossé, 1987: 87; Castoriadis, 2006: 10 and 45–358), which possibly incites Greeks to permanently seek stability, order, and unity. However, together with the desire for the latter, a clear awareness exists about the polarization, the dualism in the world and in society, which leads to classify all beings and things into a series of oppositions. This is characteristic of a military type of society that owns two historical traditions strongly opposed to one another: the Dorian, which represents the dynamic, individualistic, and centrifugal aspects; and the Ionian, an image of the static, community-based, and centripetal aspects. Added to this, and in correspondence with the previous discussion, two socially, culturally, and politically opposed groups exist: the *demos* and the aristocracy (Roche Cárcel, 2017).

Faced with that instability, the attachment to the earth arises as one of the most tenacious ways to find a certain degree of stability, since even the economic interests of city dwellers continue to lie in the countryside (Finley, 1975: 72). In this connection, agriculture becomes a task linked to economics, but also to religion, society, culture, politics, and war (Gallego, 2003: 327). Consequently, despite being an essential feature of peasant households (Garnsey, 2003: 90), individualism coexists with the awareness of the need to socialize and to exchange. That is why the origin for the success of *oikos*—the house—internally derives from a single arrangement of the family group, ownership and the residence, albeit complemented through the external links provided by the exigence of public life (Buxton, 2000: 74; Gallego, 2003: 70) and commerce.

Therefore, the latter, along with agriculture, cattle-raising, manufacturing, and mining, played a significant role in the Greek generation of wealth (Mossé, 1993: 50–58; Gallego, 2003: 200).

The translation into Spanish of *polis* as *ciudad* is insufficient, as it designates a complex human, legal, political, natural, and religious notion (Roche Cárcel, 2017). In any case, it refers to the individual and collective person, insofar as, although Greeks reveal a clearly individualistic trend, their social conception is fundamentally "communist" (Kitto, 1979: 107); hence the evolution of the *polis*—the family and the city (De Coulanges, 1986: 116 ff.)—does not depend on two forces, but on three, since that of the individual must be added to them (Glotz, 1957: 4).

Together with the polis, the great Greek creation was Democracy, which has an innate origin in the myth for being linked to the originary chaos and the essential instability of human life. But what is exactly understood by democracy, particularly the Athenian one? A suitable answer must take into account its peculiar characteristics and its fundamental principles: law and *dike*—justice; *isonomy*—equality of rights and duties before the law and political participation in the State and in power; *eleuthería*—freedom; *isogoría*—equality of birth; *isegoría*—freedom of speech; and *koinonía*—community or association, that is, what is common (Castoriadis, 1998: 117; Gil, 1989: 49; Benéitez, 2005: 41; Lévêque, 1993: 48). It is also necessary to consider the specific format adopted by the Constitution, always in progress and never closed or unmovable (Iriarte, 2010: 5), and which offsets power through a central government, an oligarchic council, and a general assembly of citizens (Tilly, 2010: 58), as well as its own institutions and the specific aspects which define citizenship and refer both to individual citizens and to the collectivity. That is why the alternative between individualism and the idea of community cannot be univocally proposed in a democracy, as both things are indissolubly linked (Hauser, 1988: 1089).

Speaking about Athenian democracy does not mean referring to equality, though, because contrary to our expectations, it is a slave-owning society, to such an extent that the most important opposition takes place between free individuals and slaves, to which must be added the antagonism existing between citizens and metics, and between the rich and the poor, as well as the one which opposed city dwellers and the people who lived in the countryside (Lévêque, 1978: 7), without forgetting the sharp distinction between men and women. As a matter of fact, the Indo-European invasions install the prevalence of the masculine principle over the feminine one in Greece, which is why the male will and ego are imposed and thrive. Greece thus devises a self-conscious type of masculine intelligence which understands the world rationally and judges it responsibly through empirical facts; and all this (Campbell, 1992: Chap. IV) for the ultimate purpose of developing and maturing the human being, and not of feeling the gods. That entails a markedly patriarchal social structure based on marriage and private property (Mossé, 2001: 150).

This explains why Mother Goddess no longer constitutes the focus of worship and why she is replaced by the Olympic masculine gods in the main pantheon. Furthermore, the old divinity undergoes a mythological defamation with Zeus' victory against Typhoon or that of Apollo over Python, which symbolizes the triumph of a new masculine justice order over the archaic, "agitated," and "chaotic" matriarchal power (Iriarte & González, 2008: 123). The influence exerted by Mother Goddess does not disappear, though, since it will survive by segmenting her competences between different goddesses that personify one of her aspects—Demeter, Hera, Aphrodite, Artemis, and Athena. Nonetheless, despite continuing to symbolize the natural life of animals, plants, and mankind, the affirmation of courage, fertility, and erotic love, these goddesses now represent the home and the bond of women with men (Dunn, 2008: 30). Not in vain, the Olympic pantheon has been organized through the adoption as a model of the family and patriarchal hierarchical structure of the new Indo-European society which is superimposed over the atavistic matriarchal one.

The Diverse Greek Creation Myths

Aristotle carries out a summary of the various old Greek theogonies (*Metaphysics*, 1091a–1092a), but, for the purpose of this chapter, my attention will especially focus on three different and contradictory paradigms about creation:

1. Homer's "world without beginning," which has no interest in the origin of the world, since it revolves around human adventures and anthropomorphic gods instead. Furthermore, the humans and divinities in the *Iliad* and the *Odyssey* have reached their full maturity and, most outstandingly, it is Man and not creation that constitutes the starting point (Boorstin, 2008: 35–42).
2. The "intuition of a void prior to being." In Hesiod's *Theogony*, creation is erotic, which entails a birth that leads to life, eternally renewed with creativity (Steiner, 2002: 45, 95 and 264). This means that it reveals innumerable acts of creation (Boorstin, 2008: 43)—an echo of Mother Goddess and of her successive restarts—and additionally refers to gods' feats and not to world genesis (Arendt, 2005: 92). More precisely, *Kháos* seems to be the matrix of everything that has been created, the origin of all things, a bottomless, vast and dark, confusing and gloomy void, and a shapeless and undifferentiated place (West, 1966: 192–193; Chantraine, 1999: 1246) and a great feminine mystery: that is, the primordial strange, the hollow, the abysmally profound other (Jung, 2015: 141). From it arose the first deity, *Gea* "the broad-breasted one," Mother Earth, the "permanently safe seat," "the ground of the world," the goddess of "solid foundations," as highlighted in *Homeric Hymn to the Earth, 1*. Therefore, Nature, Divinity, and Men emerge from the unspeakable chaos and they are all consequently "children of the abyss," unceasingly fighting against it.

   However, the fight also characterizes the whole created order and seems to be immanent to the actual generation process. It is worth noting, in this respect, that the emergence of divinities entails universal suffering, and that it takes place to fight the forces of disorder and to find—each time and definitively—the order and unity that their differentiation and individuality broke (Vernant, 2001: 41; Vernant, 2003: 15 ff). Consequently, this old mythical awareness of the impossibility to solve the problem of world disorder will appear once and again in every social and cultural activity, and at any time (Castoriadis, 2006: 10 and 45–358).
3. The search for the original unity. According to Parmenides, nothingness does not exist and it is impossible to think about it (Steiner, 2002: 35). Indeed, the Greek philosopher understands—unlike Hesiod—

that an identity exists between being and thinking, insofar as being is not nothingness and does not originate from it either, and neither does it return into its heart: "since there is being, nothingness does not exist" (*Poem* II, fragment 6, 1). Starting from this idea, Plato (*Ion*) points out that the artist is not a primary creator (Steiner, 2002: 61); instead, he defends (*Timaeus*, 30) that a chaotic substratum exists in the universe and, therefore, the demiurge—the craftsman who makes the world—knows that he cannot fully rationalize the *cosmos* that he is building. This fits in with Plato's deep mistrust in the human being, due to the outcomes derived from the process of History and to rationality, which he considers insufficient to achieve his main objective: the search for unity and for the absolute (Muñoz, 1975: 77). In contrast, the philosopher suggests (*Timaeus*) that the world is formed with pure ideas and impure immaterial substances and that, rather than a creator, there are re-creation cycles (Boorstin, 2008: 52).

The Common Features of Creativity

In addition to the specific factors that generated the creation myths dealt with here, Greeks generally assign the following features to creativity:

1. Creativity essentially follows the masculine principle, even though the feminine one survives and becomes powerfully present in certain moments and circumstances. For instance, although the *Iliad* portrays the masculine Indo-European world, "the Goddess returns" in the *Odyssey*, since Ulysses' itinerary may constitute a visionary journey to reintegrate the masculine and the feminine as well as the reciprocity of their relationship (Campbell, 2015: 221 and 247).
2. Thus, Greek creativity, dominated by the masculine principle, is determined by a deep polarization that materializes in an opposition between the collective and the individual sphere, between unity and diversity, between *logos* and irrationality, between matter and idea, and between immanence and transcendence. Greek creativity thus moves between the collective and the individual—in theatre (Kitto, 1979: 122; Duvignaud, 1966: 242) and in art (Fischer, 1986: 53; Finley,

1975: 159; Buxton, 2000: 68; Bowra, 1983: 123; Boardman, 1988: 311–341; Tatarkiewicz, 1987: 345; Blanco Freijeiro, 1990: 334).

3. Likewise, creativity reveals a deep tension between unity and diversity. Thus, artists were divided into two opposed schools already during the Archaic Age: the Dorian and the Ionian cultures, which fluctuate between fiction and truth, between imitation and creation, and between Beauty and convenience. In short, Greek Antiquity knows and appreciates a twofold beauty: that of *symmetry* and the individual one of convenience (Tatarkiewicz, 1987).
4. *Logos* and irrationality are inherent to the Greek creativity process. The noun *logos*—as stressed by Heidegger in the *Introduction to Metaphysics*—derives from the verb *légein*, which means "to speak," "to refer," "to select," "to gather," "to assemble." Therefore, all these actions suggest that *logos*—the same as creativity—refers us back to a way of thinking, counting or calculating, and relating (Bilbeny, 2012: 86). Furthermore, "the oldest word for the reign of the word conceived like this, to say, is called *logos* … the word *logos* itself, the noun for *saying*, is simultaneously used for *being*, i.e. for the presence of what is present" (Heidegger, 1987: 213), which means that language—and creativity—are home to being (Trías, 1984: 97–98).

   However, if rationality is one of the essential characteristics of creativity, so is irrationality, insofar as powerful irrational forces of an external or internal origin exist in Greece, namely: *mania*, madness, *ATE*, the fear of *Miasma* and of contamination, *Phthónos*, the envy of gods, the continued existence of demons, and, finally, the fear of freedom which characterizes the whole Hellenistic period; and the goddess of fortune *Tyché* who, with her swings, batters humans in such a way that her actions—including the creative ones—rational as they might be, carry a wild and unexpected component with themselves; they trigger unforeseen consequences (Dodds, 1983).
5. Greek art is material, and at the same time, ideal (Flynn, 2002: 33–35). In this connection, as stated by Plato (*Cratylus* 430a), both word and image must transmit the fundamental qualities of the idea underlying them (Bohme, 2000: 178–200).
6. The key aesthetic concept for Greeks is that of *mimesis* or imitation, which means that, rather than creation, this has to do with re-creation,

by means of which the artist or the philosopher reflect on what exists (Steiner, 2002: 32).

7. Creativity in Greece basically develops within space and not over time, since the interpretation of being neither knows nor understands the fundamental ontological function of time (Heidegger, 2003: 49), forgets about it in religious representations (Kerényi, 2010: 78), and limits it, so that human existence can progress inside it (Glacken, 1996: 51–52).

Greek creativity would thus arguably lie closer to the spatial category than to the temporal one (Beriain, 2008: 43), because Greeks were not expelled from paradise and deeply loved their territory; hence—like Egyptians—they were "builders of space"—so is Greek creativity; they filled the natural "voids" and urbanized nature (Beriain, 2000: 36). Nonetheless, this space devoid of time is not—as that of Mother Goddess was—absolute and infinite, but constrained by measures and limits.

3 Creation Myths as Originators of Contemporary Society

Creativity and the Cultural Continuity with the Past

After analyzing and interpreting the myths about the origin of Mother Goddess, about biblical *Genesis*, and about the old Greek myths—which correspond to three types of religiousness and to the same number of societies—and from the objectives proposed, a conceptual map of creativity has been drawn which allowed me to define the most important characters unveiled by those myths.

More specifically, it was possible for me to obtain a series of variables organized on a polar basis which set the limits of creativity and make it possible to show its different orientations toward fusion or differentiation, space or time, the vital or the spiritual sphere, the feminine or the masculine, the body or the mind, the collective or the individual sphere, security or risk, chaos or order, freedom or necessity, and creativity or mimesis. These poles continue to be valid in contemporary society, which suggests that the aspects of creativity unveiled by myths refer us back to the origin, but they are also originators.

The definition of creativity can thus be completed through a set of features characterized by ambivalence and polarization. In correspondence with what creation myths suggest, creativity may be masculine and/or feminine (Tönnies, 2011: 272–273); individual and/or collective (Weber, 1984: 241; Weber, 2007: 35–38; preliminary study by Joaquín Abellán; Elias, 1991: 74–88); physical, referring to the flesh and the body, and/or conceptual (Rifkin, 2000: 50; Maffesoli, 2004: 148); free, contingent, and/or necessary (Castoriadis, 2006: 127–130; Steiner, 2002: 134–136; Valencia López, 2016: 127; Arizpe, 2001: 39; Sánchez Capdequí, 2014: 204); safe and/or risky (Ramos Torre, 1999: 231; González, 2006: 389 and 408 ff.); creative or destructive (Steiner, 2002: 126); chaotic or organized (Morin, 1981: 76 ff.; Ramos Torre, 1999: 222–234; Ramos Torre, 2012: 70–73); merging and/or unequalizing (Maffesoli, 2004: 89–145; Ariño, 2016: 62–94); it may take place in space and/or in time (Florida, 2009: 74–213; Bergua et al., 2016: 138 ff.; Beriain, 2000: 48; Beriain, 2008: 44–46; Steiner, 2002: 79); and, finally, it is either mimesis, re-creation of the existent, or creation per se (Boorstin, 2008: 52).

Broadly speaking, those pairs either have merged or have become polarized in the three mythical stages selected, albeit not in an absolute manner. Matriarchal society, for example, is characterized by the fusion of concepts into a whole, although it has been steadily incorporating aspects related to invasive masculine divinities that have contaminated its old essences. In turn, patriarchal society follows a deeply and vehemently polarizing approach—an anti-feminine and anti-natural one. Finally, the unstable fratriarchal and democratic society, despite adopting a polarizing attitude, is not so radical in its fractures, as one of the poles becomes the dominant one, while the other assumes a subordinate role without completely disappearing.

Among postmodern tribes, there seems to be a certain revival of the Great Mother Goddess (Maffesoli, 2004: 89–145), because of her characteristic corporeality (Joas, 2013: 214), for her emotional and amorous fusion, and for her vital power, as well as for her fertile energy that creates new ways of living. Life has thus become a collective work of art (Maffesoli, 2004: 44–183), a/the great creation, which refers us back to Mother Goddess's worldview, but also to that of Father God, insofar as life overflows and transcends itself (Simmel, 2014: 1 and 24) and gives rise to moments of religious (Nocera, 2009: 101) and even revolutionary effervescence (Durkheim, 1993: 348–349). However, neither should we forget the teaching of nineteenth-century philosophical vitalism, which has reminded us that, at present—both the market and the advertising industry are well aware of it—life remains unsatisfied and restless, that it constitutes an unattainable and illusory goal, which leaves us deeply empty.

The presence of Father God also becomes visible in today's society, though. First and foremost, because, by means of creation, something emerges from the nothingness (Zweig, 2015: 14), the same that happens to the world after being created by God. The "economy of access" (Rifkin, 2000: 50), contemporary art (Guasch, 2002), and the virtuality of new technologies are rooted in the conceptualism of biblical creativity, just as the innovation of technological and financial enterprises does when their creativity causes growing inequality (Ariño, 2016: 62–94). At the same time, the spirit of Yahve is present in the ambivalent individual and collective creativity of charisma or of genius. The charisma of extraordinary figures (Weber, 1984: 241) generates a new order opposed to the traditional one, whereas genius may certainly have a feminine nature (Tönnies, 2011: 272–273), but can likewise be a male individual whose creativity is determined by the social process to such an extent that he can even be defeated by it (Elias, 1991: 74 ff.). Democracy drives freedom as well as personal and social creativity, because it represents an ongoing self-creation of social action (Dewey, 1996: 227–232; Touraine, 1995). Finally, biblical time comes to its close in the protean, random, and constructed postmodern identity, a life project (Nair & de Lucas, 1999: 77 ff.; Sennett, 2000: 140; Bauman, 2003: 88 ff.).

Finally, the Greek way of understanding creativity has left its footprint at present in the problematic conception of human nature, in unstable and fragile human beings marked by chaos in their infrastructure (Morin, 1981: 76 ff.). This fragility and instability ultimately take us back to the notions of "contingency," "uncertainty," and "risk" (González, 2006: 389 and 408 ff), as well as to lighter and weaker ties (Illouz, 2009: 205).

Likewise, Greeks' spatial construction can currently be felt in creative cities (Florida, 2009: 74–213) or in offices (Bergua et al., 2016: 38 ff), where economic growth, the exchange of news, and happiness take place. It is as if the city or the workplace became the safe and merging mother or the Paradise from which we were never expelled.

All in all, creativity currently generates a society that is simultaneously feminine and masculine, merging and differentiating, bodily and spiritual, material and conceptual, individual and collective, safe and risky, free and necessary. Life in it is full or empty and, consequently, creativity moves between being and nothingness, between the inheritance left by Mother Goddess and those of Father God and Son God. This accordingly confirms not only that cultural continuity exists but also that nothing is lost forever (Bellah, 2005: 72).

Creativity Builds the Autonomous Individual and Society

In any case, creativity is on the one hand inherent to autonomous individuals, to "legally-of-age beings"—the great Kantian enlightened ideal, not completely lost, despite its successive failures—since it produces individual, unique and unrepeatable, "original", rational and emotional, feeling and thinking beings, which is the psychological function of creativity. This explains why one's own personal life is concocted by means of creativity and freedom (the Bible, Greece, democracy), with nobody imposing the lifestyle or the actual meaning of existence on anybody else. However, an unshared, nonerotic type of creativity, not conceived as a common issue—in short, not socialized—does not seem possible, insofar as it once again reminds us that the human being either is social or is not.

This argument consequently explains why creativity performs a first-order sociological function, since it edifies personal biographies and helps found history, creates the time, is temporality, and, therefore, performs an essential historical function too.

Bibliography

Arendt, H. (2005). *La condición humana*. Paidós.

Ariño, A. (2016). *La secesión de los ricos*. Galaxia Gutenberg.

Arizpe, L. (2001). Cultura, creatividad y gobernabilidad. In *Estudios Latinoamericanos sobre Cultura y transformaciones sociales en tiempos de globalización* (pp. 31–48). CLACSO, Consejo Latinoamericano de Ciencias Sociales.

Bauman, Z. (2003). *Modernidad líquida*. FCE.

Bellah, R. N. (2005). What is axial about the axial age? *Archives Européennes de Sociologie, XLVI*(1), 69–87.

Bellah, R. N. (2017). *La religión en la evolución humana. Del Paleolítico a la era axial*. CIS.

Benéitez, B. (2005). La ciudadanía de la democracia ateniense. *Foro Interno, 5*, 37–58.

Bergua, J. A. (2015). *Postpolítica. Elogio del gentío*. Biblioteca Nueva.

Bergua, J. A., Carretero, E., Báez, J. M., & Pac, D. (2016). *Creatividad. Números e imaginarios*. CIS.

Beriain, J. (1999). Del reino de Jaungoikoa (Dios) al politeísmo moderno. In J. Beriain & R. Fernández Ubieta (Eds.), *La cuestión vasca* (pp. 70–86). Anthropos.

Beriain, J. (2000). *La lucha de los dioses en la modernidad. Del monoteísmo religioso al politeísmo cultural*. Anthropos.

Beriain, J. (2008). *Aceleración y tiranía del presente. La metamorfosis en las estructuras temporales de la modernidad*. Anthropos.

Bilbeny, N. (2012). *Ética*. Ariel.

Billinghurst, J. (2007). *Mujeres tentadoras*. Océano.

Blanco Freijeiro, A. (1990). *Arte griego*. C.S.I.C.

Boardman, J. (1988). *Arte y arquitectura griegos*. In *Historia Oxford del mundo clásico. I. Grecia* (pp. 311–353). Madrid, Alianza.

Bohme, G. (2000). *Platons theoretische Philosophie*. Metzler.

Boorstin, D. J. (2008). *Los creadores*. Crítica.

Bowra, C. M. (1983). *La Atenas de Pericles*. Alianza.

Bru, M. (1990). La mujer en el arte prehistórico. Simbología y Representación. *Actas de las Terceras Jornadas de Investigación Interdisciplinar*, 17–28.

Buxton, R. (2000). *El imaginario griego. Los contextos de la mitología*. Cambridge University Press.

Campbell, J. (1992). *Diosas*. Atalanta.

Campbell, J. (2015). *Las Máscaras de Dios: Mitología Occidental*. Alianza.

Castoriadis, C. (1998). La polis griega y la creación de la democracia. In *Los dominios del hombre. Las encrucijadas del laberinto* (pp. 97–132). GEDISA.
Castoriadis, C. (2006). *Lo que hace a Grecia. 1. De Homero a Heráclito. Seminarios 1982–1983. La creación humana II*. FCE.
Cencillo, L. (1970). *Mito. Semántica y realidad*. Biblioteca de Autores Cristianos.
Chantraine, P. (1999). *Dictionnaire étymologique de la langue grecque. Histoire des mots*. Klincksieck.
Cioran, E. M. (1986). *La caída en el tiempo*. Planeta Agostini.
Cirlot, J. E. (1988). *Diccionario de símbolos*. Editorial.
De Coulanges, F. (1986). *La ciudad antigua*. Biblioteca EDAF.
Delporte, H. (1982). *La imagen de la mujer en el arte prehistórico*. Colegio Universitario de ediciones Istmo.
Dewey, J. (1996). Democracia creativa: la tarea ante nosotros. In *Liberalismo y otros ensayos* (pp. 199–205). Edicions Alfons El Magnànim.
Dodds, E. R. (1983). *Los griegos y lo irracional*. Alianza.
Dunn, M. (2008). *Diosas. La canción de Eva. El renacimiento del culto a lo femenino*. Malsinet editor, colección Nueva mujer.
Duque, F. (2000). *Filosofía para el fin de los tiempos*. Akal.
Durkheim, É. (1993). *Las formas elementales de la vida religiosa*. Alianza.
Duvignaud, J. (1966). *Sociología del teatro. Ensayo sobre las sombras colectivas*. FCE.
Eisenstadt, S. N. (1986). The axial age breakthroughs-their characteristics and origins. In *The origins and diversity of axial age civilizations* (pp. 1–29). State University of New York.
Eliade, M. (1952). *El mito del eterno retorno*. EMECE.
Eliade, M. (1989). *Herreros y alquimistas*. Alianza.
Elias, N. (1991). *Mozart. Sociología de un genio*. Península.
Engels, F. (2013). *El origen de la familia, la propiedad privada y el Estado*. Alianza.
Finkelstein, I., & Silberman, N. A. (2005). *La Biblia desenterrada. Una nueva visión arqueológica del antiguo Israel y de los orígenes de sus textos sagrados*. Siglo XX.
Finley, M. I. (1975). *Los griegos de la Antigüedad*. Nueva Colección Labor.
Fischer, E. (1986). *La necesidad del arte*. Planeta-Agostini.
Florida, R. (2009). *Las ciudades creativas. Por qué donde vives puede ser la decisión más importante de tu vida*. Paidós.
Flynn, T. (2002). *El cuerpo en la escultura*. Akal.
Freud, S. (1985). *Tótem y Tabú*. Alianza Editorial-Club Internacional del Libro.
Frye, N. (1988). *El Gran Código*. Gedisa.

Frye, N. (1996). *Poderosas palabras. La Biblia y nuestras metáforas*. Muchnik.

Gadamer, H.-G. (1996). *La actualidad de lo bello*. Paidós.

Gallego, J. (2003). Comunidad aldeana y sociabilidad campesina en la Grecia Antigua. In J. Gallego (Ed.), *El mundo rural en la Grecia antigua* (pp. 327–380). Akal.

Gamble, C. (2001). *Las sociedades paleolíticas de Europa*. Ariel.

Garnsey, P. (2003). El campesinado: subsistencia y supervivencia. In J. Gallego (Ed.), *El mundo rural en la Grecia antigua* (pp. 71–102). Akal.

Gil, L. (1989). La ideología de la democracia ateniense. In *Cuadernos de Filología Clásica*, No. 23, 30–50. Universidad Complutense.

Gimbutas, M. (1991). *Diosas y dioses de la vieja Europa*. Ediciones Istmo.

Gimbutas, M. (1997). La religiosidad de la Diosa en la Europa Mediterránea. In *Tratado de Antropología de lo sagrado: las civilizaciones del Mediterráneo y lo Sagrado* (pp. 41–62). Trotta.

Glacken, C. J. (1996). *Huellas en la playa de Rodas. Naturaleza y cultura en el pensamiento occidental desde la Antigüedad hasta finales del siglo XVIII*. Ediciones del Serbal.

Glotz, G. (1957). *La ciudad griega*. Unión tipográfica Hispano Americana.

González, J. M. (2006). *La diosa Fortuna. Metamorfosis de una metáfora política*. Antonio Machado libros.

Guasch, A. M. (2002). *El arte último del siglo XX. Del posminimalismo a lo multicultural*. Alianza.

Guilaine, J., & Zammit, J. (2002). *El camino de la guerra. La violencia en la prehistoria*. Ariel.

Hauser, A. (1988). *Historia social del arte y de la literatura*. Labor.

Hayoun, M. R. (1995). El judaísmo. In J. Delumeau (Ed.), *El hecho religioso. Enciclopedia de las grandes religiones* (pp. 201–252). Alianza.

Heidegger, M. (1987). *De camino al habla*. Delós.

Heidegger, M. (2003). *Ser y tiempo*. Trotta.

Husain, S. (2001). *La Diosa. Creación, fertilidad y abundancia. Mitos y arquetipos femeninos*. Taschen.

Illouz, E. (2009). *El consumo de la utopía romántica. El amor y las contradicciones culturales del capitalismo*. Katz.

Iriarte, A. (2010). Atenas o el arte de lo político. *Revista Bitarte, 50*, 29–42.

Iriarte, A., & González, M. (2008). *Entre Ares y Afrodita. Violencia del erotismo y erótica de la violencia en la Grecia Antigua*. Abada editores.

Joas, H. (2013). *La creatividad de la acción*. CIS.

Jung, C. G. (2015). *Arquetipos e inconsciente colectivo*. Paidós.

Jünger, E. (1998). *El libro del reloj de arena*. Tusquets.

Kerényi, K. (2010). *Imágenes primigenias de la religión griega. II. Hermes el conductor de almas. El mitologema del origen de la vida masculina*. editorial Sexto Piso.

Kitto, H. D. F. (1979). *Los griegos*. EUDEBA.

Knöbl, W. (2017). El debate de 'las modernidades múltiples, su fundamento sociológico-religioso y sus escollos. In C. Sánchez Capdequí (Ed.), *La creatividad social: narrativas de un concepto actual* (pp. 263–315). CIS.

Lessing, E., & Sollers, P. (1994). *Mujeres, mitologías*. M Moleiro editor.

Lévêque, P. (1978). Las diferenciaciones sociales en el seno de la democracia ateniense del siglo V. In D. Roche (Ed.), *Órdenes, estamentos y clases*. Siglo XXI.

Lévêque, P. (1993). El papel de la religión en la génesis de las ciudades. *Revista de Occidente, 143*, 43–60.

Lucie-Smith, E. (1994). *La sexualidad en el arte occidental*. Destino.

Maffesoli, M. (2004). *El tiempo de las tribus. El ocaso del individualismo en las sociedades posmodernas*. Siglo XXI.

Morin, E. (1981). *El método I. La naturaleza de la naturaleza*. Cátedra.

Mossé, C. (1987). *Historia de una democracia: Atenas*. Akal.

Mossé, C. (1993). El hombre y la economía. In J.-P. Vernant (Ed.), *El hombre griego*. Alianza.

Mossé, C. (2001). *La mujer en la Grecia clásica*. Nerea.

Mumford, L. (1979). *La ciudad en la historia*. Infinito.

Muñoz, I. (1975). La ideología de la aristocracia griega antigua. *Revista de Estudios Políticos, 199*, 65–82.

Nair, S., & de Lucas, J. (1999). *Inmigrantes. El desplazamiento en el mundo*. Ministerio de Trabajo y Asuntos Sociales. Secretaría General de Asuntos Sociales. Instituto de Migraciones y Servicios Sociales.

Nocera, P. (2009). Los usos del concepto de efervescencia y la dinámica de las representaciones colectivas en la sociología durkheimiana. *REIS, 127*, 93–119.

Ortiz-Osés, A. (1993). *Las claves simbólicas de nuestra cultura: Matriarcalismo, patriarcalismo y fratriarcalismo*. Anthropos.

Ramos Torre, R. (1999). *Red, identidad, espacio y tiempo* (p. 86). *Revista Española de Investigaciones Sociológicas*.

Ramos Torre, R. (2012). Los avatares de lo trágico en la sociología: de Tocqueville a Luhmann. In J. A. Roche Cárcel (Ed.), *La Sociología como una de las Bellas Artes. La influencia de la literatura y de las artes en el pensamiento sociológico*. Anthropos.

Ries, J. (1997). *Tratado de antropología de lo sagrado: las civilizaciones del Mediterráneo y lo sagrado*. Trotta.
Rifkin, J. (2000). *La era del acceso. La revolución de la nueva economía*. Paidós.
Roche Cárcel, J. A. (2017). *Entre el Monte de Apolo y la vid de Dioniso. Naturaleza, Dioses y Sociedad en la arquitectura teatral de la Grecia Antigua*. Anthropos.
Rodríguez Adrados, F. (1983). *La Democracia ateniense*. Alianza.
Sánchez Capdequí, C. (2014). *El dinamismo de los valores. Crisis y creatividad en la sociedad moderna*. Anthropos.
Sennett, R. (2000). *La corrosión del carácter. Las consecuencias personales del trabajo en el nuevo capitalismo*. Anagrama.
Simmel, G. (2014). *Intuición de la vida*. Prometeo.
Smith, P. (2003). *Arte rupestre paleolítico en Cantabria*. Asociación Cántabra para la defensa del patrimonio Subterráneo.
Steiner, G. (2002). *Gramáticas de la creación*. Siruela.
Tatarkiewicz, W. (1987). *Historia de la Estética. I. La estética antigua*. Akal.
Tilly, C. (2010). *Democracia*. Akal.
Tönnies, F. (2011). *Comunidad y asociación. El comunismo y el socialismo como formas de vida social*. Biblioteca Nueva-Minerva.
Touraine, A. (1995). *Producción de la Sociedad*. Instituto de Investigaciones Sociales UNAM-IFAL.
Trías, E. (1984). *Drama e identidad*. Ariel.
Valencia López, H. (2016). La política y la democracia como creaciones imaginarias: de los griegos a nosotros. *Atenea* 513, I Sem., 125–135.
Veblen, T. (2008). *Teoría de la clase ociosa*. Alianza.
Vernant, J.-P. (2001). *El individuo, la muerte y el amor en la Antigua Grecia*. Paidós.
Vernant, J.-P. (2003). *El universo, los dioses, los hombres. El relato de los mitos griegos*. Anagrama.
Weber, M. (1984). *Economía y Sociedad. Esbozo de sociología comprensiva*. FCE.
Weber, M. (2007). *Sociología del poder. Los tipos de dominación*. Alianza.
West, M. L. (1966). *Hesiod Theogony*. Oxford University Press.
Zweig, S. (2015). *El misterio de la creación artística*. Sequitur.

3

Creativity: Sociocultural Order and Chaos, Memory of the Past, and Projection of the Future

1 Myth and Rite—The Same as Creativity—Turn Disorder into an Ingredient of Order

The order–chaos interaction arises as an important factor within the creative time process that I have been referring to earlier, as shown by the origin myths and Science, which addresses the generation of life. As far as creation myths are concerned, traditional societies cause little disorder and, for that reason, strongly resist historical evolution. Thus, they repeat the old order, reproduce themselves without significant variations, and do not generate an "orderless" future. Furthermore, if a place exists for disorder in these societies, it is because disorder is feared and immediately translated into order. And when the latter becomes transmuted into its opposite, that does not imply its collapse or disappearance, but its reinforcement or reconstitution under a new shape. In any case, evil, death, and chaos come from outside, from elsewhere, since they result from indistinctness, from uncertain limits, from the poorly defined boundaries of two different universes—the wild nature and the organized city—and from confusing categories of good and evil, the sacred and the profane, for instance. It is true, though, that paths open between both worlds

J. A. Roche Cárcel, *Creativity and Time: A Sociological Exploration*, Palgrave Studies in Creativity and Culture, https://doi.org/10.1007/978-3-030-84838-5_3

which connect them, thresholds or spaces of transition that must be crossed through tests or rites of passage (Balandier, 2014, pp. 35–181).

On the other hand, for the creation myths of these traditional societies, it is an initial disorder that engenders order. In effect, in the beginning there was chaos, an initiatory temporality which leads us outside temporality, or even to a cancellation of historical time. Expressed differently, it is a moment when nothing exists, when all the things, the natural spaces and beings must be created, and when humans turn hope into a break of the established order, transforming the present time lived, similar to chaos and evil, into a future that brings a different, yearned-for order. After all, this time specifically implies the birth of the world or of the expectation of a new society. However, nothing that has been created remains stable; and its permanence is not guaranteed either, since all that has been engendered is doomed to degradation in the long run. Indeed, social forces deteriorate, society wears out, and the cosmos generates its own decay insofar as "energy becomes exhausted in the heat of life and time breaks up to such an extent that it entails the end of the future" (Balandier, 2014, pp. 19–24).

Alongside this original myth, there is another which narrates the new beginnings differently, as the latter emphasizes the break with the currently accepted History, for the purpose of establishing another different and yearned-for history which represents an ideal stage. Hence, this type of myth is framed within society's historical time, rather than within the temporality which precedes time and, in such a myth, order—harmony—becomes projected toward the future, and not toward the past. What is more, the world announced by it appears as true, as an institutionalizing order which relinquishes the pristine return that characterizes the other creation myth modality (Balandier, 2014, p. 25).

Some examples of past origin myths (Bellah, 2017, p. 300, 373 and 619) confirm the insights provided earlier. By way of example, both consciences—that of order and that of disorder—were present in ancient Egypt. Chaos was conceived as "complete darkness" and "limitless water," the predecessor from which the first god emerges. Moreover, it surrounded the finite universe, or rather, constantly pervaded it, a persistent human action being needed to control it. And finally, when the cosmos aged, it survived that cosmos and became reabsorbed in it. In turn, creation within Vedic thinking was essentially chaotic, disorganized and immature and, therefore, imperfect and unfinished, whereas the Ugarit myths about world creation portray the supreme god *El* managing to overcome the chaos of waters, seas, and rivers—the symbol of flow and instability.

According to Christianity—as shown in the preceding chapter—God firstly and paradoxically generates the material primary elements: "In the beginning, Elohim created the heavens and the earth" (Genesis 1,1). Before the first day of creation, there was a shapeless, disorganized mass without plant life, covered by waters and submerged in darkness, on which the God's Spirit landed. Only confusion and disorder reigned in this universe prior to the creation of an organized world, a distant echo of the old Mother Goddess, through which the Creator hovered.

The myths about the origin in African traditions succeed in making society think about itself and legitimize its created order. Not in vain, they explain how this order is full of risks, successive failures, and reconstructions, since it must always be rebuilt (Balandier, 2014, p. 22).

According to Hesiod's Greek creation myth, *Kháos* arises as the matrix of everything that has been created, the origin of all things, a bottomless void, vast and dark, confusing and gloomy, a shapeless and undifferentiated place, as well as a great feminine mystery, that is, the primordial strange, the hollow, the abysmally deep other. The first deity, Gaia, arose from it, "the soil of the world," so that Nature, Divinity, and Humans come from the unmentionable chaos and all of them are consequently "children of the abyss," despite unceasingly fighting against it.

Therefore, this old mythical awareness about the impossibility to solve world disorder will once again appear in every social and cultural activity, as well as in every subsequent period of Western civilization (Castoriadis, 2006, p. 10 and 45–358); for instance, with the Epicureans, who apprehend the consequences of an order placed above a disorder background (Balandier, 2014, p. 43). Likewise, when it comes to contemporary society, the Greek way of understanding creativity has left an imprint on the problematic conception of human nature, on the instability and fragility of human beings who are marked by chaos in their infrastructure.

Ritual clearly precedes Myth, although the former is stable whereas the latter has a more varied nature (Bellah, 2017, p. 191). Moreover, despite being related to the ritual, the myth has its own logic, pursues a specific purpose, and demands being effective. In this regard, its main function consists in integrating individuals into the heart of society and culture, and initiating them into sacred, political, and social affairs. Thus, more precisely—thanks to the rite—the course of individuals' lives goes through its most important stages between birth and death and, at the end of life evolution, they end up becoming social human beings. This means that they "enter"[1] an order, their own society, and, accordingly, that the rite works for order, that it is order in itself. After all, it is structured and structures participants in it, while simultaneously being linked to the norm and creating moderation. Neither can we forget that it serves as the instrument of natural and social regularities—of the cycles of nature and of agricultural activities—that it relates to others, in addition to acting as a corrective factor for lacks of order and as a harmonizer for human beings, both between themselves and with the world. As a corollary, rite transcends disorder; it temporarily transforms conflicts, disruptions, and diseases, additionally transforming the energies of such negative situations into positive ones. That is why the ritual builds a controlled environment, which represents the way things should be, in conscious tension with how they really are (Balandier, 2014, pp. 28–34; Bellah, 2017, pp. 189–190).

In sum, it is not gratuitous to state that the creative action of myth establishes a better accomplished order, a regenerated universe. Nevertheless, if that is what brings these two types of creation myths together, the difference between them lies in the fact that one of them refers us back to the past, while the other is projected toward the future. They both consequently represent the **original** collective knowledge[2] linked to space, as depicted by E. Cassirer (1989, pp. 52–3); it needs to be borne in mind—from my perspective—that it is original in the twofold meaning of "origin" and "originator."

Furthermore, by expressing the secret of the origin and offering an answer—a significant unity logic to that pristine order in the world of things and humans—the myth structures the sensitive universe and

[1] Also in sexual terms, because they enter adult life.

[2] Bold (emphasis) added.

endows it with meaning, without forgetting that it symbolizes ambiguity and the random nature of the social sphere. In this respect, mythical narration shows the ongoing action of opposed powers which are confronted with one another in the human being, as well as in the entire creation and its related and constant restorations of equilibrium. Certainly, after creation, those contrary forces fight for the world in an endless combat that results in full order never being achieved. Nonetheless, this tragic painful reality is considered essential, since social prosperity has to do with maintaining an eternal harmony, whereas chaos is regarded as a civilizing ferment. It follows from this discussion that myth becomes the fruit of a necessary oscillation between alliance and confrontation, and that it unceasingly ends up imposing the double play of order and chaos forces. More precisely, they act showing both the way in which disorder favors order and how the latter permanently tries to dominate or tame the former. It thus comes as no surprise that mythological comments never have an end, that mythological thinking constantly nourishes the imaginary (Balandier, 2014, p. 17–36 and 226), and, consequently, that myth constitutes one of the foundations of creativity.

Together with myth, rite practices also dialectically interrelate order and disorder, both of them conceived as inseparable from human beings' lives and from their historical evolution. What is more, in the specific case of explicitly political rites, this always unfinished play between order and disorder becomes more strongly evident, since no society can be freed from the presence of chaos. And, because chaos cannot be removed, the rite tries to act cunningly before it (Balandier, 2014, pp. 30–4). It is my understanding that such cunning play when facing disorder is nothing but creativity, an issue that will be addressed in the next chapter when dealing with play and creativity.

2 The Intricate Interrelation of Chaos and Order According to Contemporary Science

Our time is intellectually characterized by the profound acceptance of science (Watson, 2002, p. 14), which has currently become the "infratexture of infrastructures" (the deepest framework) of society (Morin, 1984, pp. 16–7). In this sense, it permits to adequately understand the interactions existing therein between chaos and order, together with the implications that they entail for time and creativity. Scientific knowledge has undoubtedly helped understand the uncertainty of matter, additionally causing a social impact too, since it has clarified complexity better than simplicity and stability, which have ended up being the exception rather than the rule. However, this increase of complexity is inseparably linked to that of disorder (Balandier, 2014, p. 57) and to its entangled dialectics with order. Not in vain—as suggested by E. Morin—there is now a more complex relationship of the previous determinism, the traditional one, which went from order to disorder, toward another, the contemporary one, in which the previous one is not banished, but where we do play with it, in such a way that the movement goes from disorder to order, and vice versa (Rivero, 2002, p. 110). As a result, order becomes relative and relational, disorder acquiring an uncertain nature as well (Morin, 1998, pp. 421–446). What is more, science has evolved from finding Newtonian harmony to identifying the order hidden in chaos (Balandier, 2014, p. 56).

In this respect, it is worth specifying the aspects that contemporary science has unveiled in relation to the chaos–order dialectics, to the link of order with time and events, as well as to the connection of it all with creativity. On the first issue, life is known to be built through a network of entropy increase processes; more specifically—as will be elaborated upon shortly—it is a process fed by the low entropy of food; in other words, it arises as a self-structured disruption, like that of the rest of the universe. Indeed, the whole cosmic evolution consists in a gradual disorder process and, accordingly, in an increased entropy; or, expressed differently, it is not—as has been usually suggested—the energy that moves the world, that drives the great history of the cosmos (Rovelli, 2018, pp. 119–123).

The Nobel Prize winner for physics E. Schrödinger reflects on this issue in *What Is Life?* (2017, pp. 54–124) and argues that, in effect, the natural tendency of things is to head toward disorder. He later clarifies, though, that life seems to be an orderly and regulated behavior of matter which not only relies on its proneness to pass from order to disorder, but largely in the preservation of the former. This means that living matter has a high level of order which requires very low entropy and, therefore, that it prevents entropic degradation by seeking equilibrium. How does it achieve that? By feeding on "negative entropy" or, to put it in another way, through the generation of order from order, extracting it from the environment, from sunlight, for example, or from proteins—particularly enzymes—as claimed by J. Monod in *Chance and Necessity* (2007, pp. 22–70 and 121–129). As a matter of fact, according to this Nobel Prize Winner for Physiology or Medicine, they are precisely the ones which carry out the creative function of order, the molecular agents behind the "teleonomic" performances of living beings. The "teleonomic" nature of the latter is defined because of their status as objects in whose structures and in the fulfillment of whose performances there is a stipulated project, an oriented, coherent, and constructive activity. After all, the selection operates at a macroscopic scale, that of the organism, which explains why the whole system is conservative, closed on itself, and unable to receive any instructions from the outside world.

In any case, although this maintaining machinery is apparently perfect, it undergoes disturbances generated in the macroscopic context that alter its structure in a gradual, but infallible way. These alterations, which have an accidental nature and happen randomly, are the ones which lie at the origin of every novelty and creation in the biosphere,[3] at the same time that chance is captured, kept, and reproduced by the machinery of invariance and thus transformed into order, rule, and necessity. This means that the random sequence of each protein is reproduced millions of times within each organism, cell, or generation by a device which

[3] Simulation—one of the key processes in evolution, together with replication and selection—also becomes the instrument of discovery and creation (Monod, 2007, pp. 150–163).

guarantees the survival of structures. More precisely, once recorded in DNA, the unforeseen singular accident is mechanically and faithfully replicated, and then translated into millions of copies, which additionally explains why we move from the realm of pure chance to that of necessity (Monod, 2007, pp. 107–122).

Consequently, entropy becomes order from chaos, and the opposite, disorder arising from order. Instead, chance turns into necessity—in a certain manner—and vice versa. However, all this is specified to a somewhat greater extent by the science of complexity, as E. Morin would call it, since physical systems show how disorder is replaced by random orders, insofar as simpler, low-degree orders exist at one end, with chaos and chance orders—of an infinite degree, the limit case—at the other. In the middle, we can find other highly subtle types of orders which may or may not be low-degree and chaotic (Bohm & Peat, 2007, pp. 151–169).

Therefore, as has been seen, biologists, geneticists, and physicists contrast but simultaneously link need and chance, order and disorder.[4] The latter in turn acts as the generator of order, which has become a mere accident—an event which offers endless possibilities and an inexhaustible fertility, though. However, and there lies the entangled dialectics that I have been already referring to, disorder acquires a creative role when it provokes a loss and a gain of order at the same time, which ultimately forges a new order that replaces the old one and can even outperform it (Balandier, 2014, p. 44).

Along with the order–chaos dialectics, science has also addressed the association of order with time. Quantum physics, for instance, has revealed that everything is made up of atoms, that their substance is shaped by an extremely small nucleus, and that they are mostly empty space; hence, despite its solid appearance, matter becomes something basically insubstantial, and the known world turns into a "spectrum." In this regard, unlike what happened before 1900, when physics provided a solution to predict the future with absolute certainty, quantum physics relied on uncertainty, insofar as the microscopic world differs from the everyday one, because it is governed by the unpredictable chance, by a

[4] Furthermore, the sets of elements in one and in the other refer us back to the relationship between the whole and its parts, between the unit and the multiplicity (Balandier, 2014, p. 44).

lack of motives. The universe is thus characterized by disorder and chance, even though not everything is chaos, since unpredictability can be predicted (Chown, 2007, pp. 3–21), as will be shown.

On the other hand, it has just been explained that living beings depend on a fundamental mechanism of molecular conservation. This means that evolution is not one of their properties and that the principle of life does not consist in evolution (as scholars have suggested since Darwin's times), but has the actual imperfections of the conservation mechanism at its root. For that reason, evolution in the biosphere implies a necessarily irreversible process which defines a direction in time, the same which characterizes the second principle of thermodynamics, the law of entropy increase.[5] And that explains why selective evolution constitutes a sort of machine to travel back in time which can show the overall upward tendency in the evolution, the improvement, and the enrichment of the teleonomic apparatus. Nonetheless, it must likewise be remembered that, within this upward evolution, cultural behavior, as well as the cultural features and the society that builds them, is—for human beings, unlike what happens with other living beings—the one orienting selective pressure, the one which presses genome evolution (Monod, 2007, pp. 126–169), and the one behind the travel back in time.

At this point, my aim consists in finding out what type of time this travel back ultimately leads to. Science apparently responds that we find ourselves in a "timeless world" where the unicity, the direction, the independence, the continuity, and the inner difference between past and future have been lost in its basic situations. Furthermore, in the universe, the "now" and the present mean nothing, or expressed differently, we are witnessing the end of the present. Not in vain, since time represents the trace of movement, as nothing moves, there is no time. On the other hand, in accordance with quantum physics, time does not show a single face, but a diverse one, by virtue of the different duration of each trajectory, because it elapses at different paces in relation to the place and the speed. Neither does it have an orientation, insofar as it happens in a dynamic field, which jumps and oscillates, which is not defined under a

[5] Therefore, the irreversibility of evolution in the biosphere makes it clear that the first and second principles of thermodynamics go hand in hand.

minimum scale, and which only becomes specified in interaction (Rovelli, 2018, pp. 33–72 and ff).

But who are the agents in this interaction? The events which occur and not the things, and which, despite not having an ontology, are the ones that build the world. Therefore, since it is made up of events, by processes, and in short, by something that takes place, the world is characterized by constituting a network of events, as well as by incessant change and transitoriness, by a temporal order in which everything is changeable and passing (Bohm & Peat, 2007, pp. 125–6; Rovelli, 2018, pp. 75–6), by being immature and, ultimately, by being continually rejuvenated.

In sum, regarding the tight connection existing between order and chaos, science has discovered that life tries to go back in time, to stop the inevitable entropy or decay of matter, feeding it with the environmental order, that is, with oxygen, light, or enzymes. However, it can eventually be inferred that this travel back leads to a sort of return to the very first moment, prior to the 0.00003 seconds after the Big Bang (Hawking, 2005, p. 168), to time 0 of cosmos birth, that is, to a timeless, disoriented, polyrhythmic, plural, jumpy—playful—chaotic and immature time, ready to explode and generate the origin of the cosmos and to create life. Once the latter arose, it did nothing but invariably replicate itself since then, even though it was hit by accidents, by the vicissitudes of events—quickly ingested by order. In turn, the latter renourishes life again, and, with it, creation. However, creation comes as a result of the parallel gain and loss of order, and—the same as life itself—it congenitally owns a teleonomic project, since it consists in an oriented, coherent, and constructive "creative" activity.

And what about social creativity? This is going to be my main concern in the following subsection.

3 Late Modern Society or Order as a Particular Case of Disorder

As shown in the preceding pages, when they clarify their places, their limits, and their boundaries, as well as the possible paths of transit between them, traditional societies have at their disposal a neat cartography or order and disorder. According to contemporary science, they are interrelated in such an entangled manner that they do not permit an easy understanding, but rather a complex one, of the territories across which they move. Likewise, in the opinion of the social scientists—and coincidences with scientific thinking become evident therein—belonging to late modern society, although chaos and order need each other (Morin, 1998, pp. 421–446), insofar as reproduction develops thanks to a constant integration of disintegration and reintegration (Luhmann 2017, p. 237), of deconstruction and construction, it only has changeable and imprecise routes. After all, it is poorly structured, it is not organized with well-established categories and does not suitably define social classifications, which entails that History walks in the shadows, that complexity increases, and that possibilities and uncertainties multiply. Society consequently perceives itself as an "improbable order" and produces a "shattered order" (Balandier, 2014, pp. 148–215).

As a matter of fact, by emptying and hiding tradition, the ties with the past (Balandier, 2014, p. 232), and by forgetting about the redeeming future (Marramao, 1989, 80 ff.), society closely relates the categories of time and order, albeit generating an order without duration—and falsely, without legitimacy. That explains why contemporary society extols the present, contracted and converted into an omnipresent and tyrannical one (Beriain, 2008, pp. 49–188), and why it is characterized by a fluidity which becomes even more volatile, immaterial, weightless, and aerial than in modernity (Sánchez Capdequí, 2004, p. 300), and by widespread change, turned into a way of life (Farson, 1973, p. 126), into an imperative and an end in itself (Bauman, 2004, p. 53). To this must be added a dominant space–time discontinuity (Koselleck, 2016, p. 130) and the awareness of an accelerated temporal evanescence—Ch. Baudelaire (2004, pp. 91–2); W. Benjamin (Picó, 1988, pp. 21–3)—together with the prevalence of the event, understood as a figure of disorder (Balandier, 2014, p. 176) and the triumph of evolution (Baumer, 1985, 379 ff.). With it arises the contingent (Giddens and Luhmann, 1996), the uncertain and the risky (González, 2006, p. 389 and 408 ff.), the passing, the transitory, the elusive, the ephemeral (Thiebaut, 1996, 313 ff.; Shattuck, 1991, p. 292), the instantaneous (Bauman, 2003, pp. 21–122)—eternalized (Maffesoli, 2020, 55 ff.)—the unstable, the disperse, the rupturist, the unexpected, the unprecedented, the accidental, the uncontrolled, the precarious, the unforeseeable, the random, the unknown, and the new: "Modernity is movement plus uncertainty."[6]

Finally, it all results in current societies being defined by an acceleration in the rhythm of social interactions (Simmel, 1986, 247 ff.), which become more immediate (Picó, 1988, p. 19), by the unanchoring of social systems (Giddens, 1993, p. 28), by the construction of an unstable social base (Balandier, 2014, p. 148) and—as had already been pointed out—by softer and more fragile personal bonds.

Summing up, order becomes ambiguous in late modern society, since—unlike what used to happen in traditional societies—it has stopped being the guiding pole to be transformed into a particular case of disorder. This justifies why, whereas some social scientists claim that the past, the archaic, and tradition refer us back to order and emerge as a source of guidance for analyzing, interpreting, establishing, and building modernity, others believe that disorder is creative because it breaks the rigidity of order and makes it possible for society to free itself and to move forward, as well as to fruitfully build a transition period in which the new can germinate. In any case, the awareness of disorder has become intensified; determinism is more limited; freedom and chance have been enlarged; and, most importantly, creative capacity has received a vigorous reaffirmation (Morin, 1998, pp. 421–446; Balandier, 2014, pp. 11–229), even though that capacity—just like the society from which it emanates—is now more uncertain than ever before.

[6] For G. Balandier (2014, p. 156), "[t]hinking this time and about this time necessarily means thinking movement."

4 The Time of Chaotic Social Events and Creativity

Creativity—Like Myth and Rite—Transforms Disorder into an Ingredient of Order or Deviates It Toward the Imaginary

As has been already mentioned, creativity generates something original only when it undertakes a process, a time narrative, which, from the present, becomes inseparably linked to the past and the future. Analyzing the old myths and rites about the creation of the cosmos, the world, and human beings, as well as contemporary science, which has addressed the creation and evolution of life, has largely helped me reach this conclusion. I also found the interpretation of both types of knowledge—that of Myth and that of Science—regarding creation through an interaction of vital and sociocultural order and chaos very useful.

When it comes to Myth, any creative act arises from chaos, and the latter always returns, which is why it must restore order over and over again; hence creativity is pertinaciously incomplete, imperfect, and—the same as the myths of the Vedas and of contemporary science—immature, young, or rejuvenating. In any case, creativity, like myth and rite, works to turn chaos into an ingredient of order or to deviate it toward the spaces of the imaginary (Balandier, 2014, p. 35), which of course does not mean that it will eventually defeat it, that it is going to forget that order always goes hand in hand with disorder, and that the latter may unexpectedly return at any time, even at the very act of creation.

Creativity—Like Life—Is Fueled by External Order, Sociocultural in This Case

In turn, the scientific narrative of life creation does not apparently differ too much from that contained in the creation myths mentioned in this chapter. With regard to that, it does not seem unwarranted to state that Science, the key entrepreneur and the original framework within which society and its future take shape, has currently begun to be reunited with tradition and myth, both from which it borrows the metaphors and symbols used in the contact with nature. Thus, the hidden, long-disappeared roots return, which explains why the West has a new passion, satisfied by means of a return to its origins (Balandier, 2014, p. 41 and 153). For this same reason, the argument most widely accepted by the scientific world about the creation and destruction of the universe, the Big Bang, and the "Great Explosion" (Longair, 1998, p. 136), the final deflagration of the cosmos, resembles the Christian teleology exposed in the Bible, with the books of *Genesis* and *Apocalypse*.

On the other hand, the idea suggested by science according to which life feeds on the order which comes from order, from the natural environment, can be extended to creativity itself, insofar as it must be remembered that creativity is the daughter of life. In this sense, from the perspective of science, creativity in life and life in creativity go together, since a generative order exists at the most profound level which is common to life, to inanimate matter, to mathematics, and to every area of experience—of human creativity—science, society, and knowledge (Bohm & Peat, 2007, pp. 1767–254). In Schopenhauer as well as in Nietzsche, the concept of will and life are closely intertwined with that of creativity (Joas, 2013, p. 182) and the psychoanalyst D. W. Winnicott (2013) frequently uses the expression "creative living" in his book *Playing and Reality*. G. Simmel in turn stresses that the flow of life combats "the historical fossilization and the formal numbness of a given cultural content, thus triggering the internal drive of cultural change." This allows life to transcend itself, and ensures that it will not stop fostering its development, in such a way that transcendence is immanent to life (Simmel, 2017, pp. 220–223).

Therefore, creativity—the same as the actual life from which it stems—is also nourished by external order, though in this case, it is basically sociocultural rather than natural. It deserves to be highlighted in this respect—following the general thesis exposed in this book—that every creative process should thoroughly search where the necessary order hides behind the chaotic social environment. And this may happen in History, in the past, in the origin—an essential source for building a coherent, originating project—that is to say, a narrative grounded in the past but projected toward the future.

Creativity and the Metamorphosis of a Present Contracted by Another Dilated One

Similarly, the order–chaos dialectics present both in the old myths and rites and in contemporary science permits to deeply reflect on the time of disorganized events, on the contemporary crisis, and on its connections with creativity. Perhaps for that reason it seems advisable to insist on the fact that the search for originality in creativity takes us back to its original nature, dialectically understood as a return to the origin of the past and as the originating element which is programmed into the future. In fact, forgetting this would mean that creativity would not be socially transcendent; it would not entail a truly structural change and might run the risk of becoming—like the actual society where it originates—liquid, contingent, unstable, ephemeral, without a past that can legitimize it and, in short, without a future.

Nonetheless, if social scientists are right and nothing disappears, historical backgrounds remain present in late modernity. Therefore, in the event that the creative process should really be free from the determination of a present that is forgetful about the past and the future and exclusively flow toward itself, then it might be submerged "under the troubled surface of change" (Balandier, 2014, p. 232) in that deep original magma. Hence, creativity could in this way give rise to that emotional energy which comes from the preterit, that revitalizing power which would succeed in renewing the hopes in the future, and, with it, in expanding the current present affected by contraction.

Bibliography

Balandier, G. (2014). *El desorden. La teoría del caos y las ciencias sociales*. GEDISA.

Baudelaire, C. (2004). *El pintor de la vida moderna*. Colegio Oficial de Aparejadores y arquitectos Técnicos, Consejería de Educación y Cultura y Fundación CAJAMURCIA.

Bauman, Z. (2003). *Modernidad líquida*. FCE.
Bauman, Z. (2004). *La sociedad sitiada*. FCE.
Baumer, F. L. (1985). *El pensamiento europeo moderno. Continuidad y cambio en las ideas, 1600–1959*. FCE.
Bellah, R. N. (2017). *La religión en la evolución humana. Del Paleolítico a la era axial*. CIS.
Beriain, J. (2008). *Aceleración y tiranía del presente. La metamorfosis en las estructuras temporales de la modernidad*. Anthropos.
Bohm, D., & Peat, F. D. (2007). *Ciencia, orden y creatividad. Las raíces creativas de la ciencia y la vida*. Kairós.
Cassirer, E. (1989). *Esencia y efecto del concepto de símbolo*. FCE.
Castoriadis, C. (2006). *Lo que hace a Grecia. 1. De Homero a Heráclito. Seminarios 1982–1983. La creación humana II*. FCE.
Chown, M. (2007). *El zoo cuántico. Guía turística del interminable Universo*. La liebre de marzo.
Farson, R. E. (1973). Las potencialidades humanas. In VVAA (Ed.), *Ensayos sobre el Apocalipsis*. Kairós.
Giddens, A. (1993). *Consecuencias de la Modernidad*. Alianza.
Giddens, A., & Luhmann, N. (1996). *Las consecuencias perversas de la modernidad: modernidad, contingencia y riesgo* (Vol. 12). Anthropos Editorial.
González, J. M. (2006). *La diosa Fortuna. Metamorfosis de una metáfora política*. Antonio Machado libros.
Hawking, S. (2005). *El universo en una cáscara de nuez*. Barcelona: Crítica.
Joas, H. (2013). *La creatividad de la acción*. CIS.
Koselleck, R. (2016). *historia/Historia*. Trotta.
Luhmann, N. (2017). La autopoiesis de los sistemas sociales. In C. Sánchez Capdequí (Ed.), *La creatividad social: narrativas de un concepto actual* (pp. 229–249). CIS.
Longair, M. S. (1998). *La evolución de nuestro universo*. Cambridge University Press.
Maffesoli, M. (2020). De L'Universalisme à l'uni-diversité postmoderne. In C. Sánchez Capdequí & J. A. Roche Cárcel (Eds.), *Modernidades regresivas*. CIS.
Marramao, G. (1989). *Poder y secularización*. Península.
Monod, J. (2007). *El azar y la necesidad*. Tusquets.
Morin, E. (1984). *Ciencia con consciencia*. Anthropos.
Morin, E. (1998). Epistemología de la complejidad. In D. F. Schnitman (Ed.), *Nuevos paradigmas, cultura y subjetividad* (pp. 421–446). Paidós.

Picó, J. (1988). Introducción. *Modernidad y Posmodernidad*. Alianza.
Rivero, C. I. (2002). El aporte de Edgar Morín al pensamiento social contemporáneo, desde una epistemología de la complejidad. *Salud de los Trabajadores*/ Volumen 10, Edición Especial No. 1 & 2.
Rovelli, C. (2018). *El orden del tiempo*. Anagrama.
Sánchez Capdequí, C. (2004). *Las máscaras del dinero. El simbolismo social de la riqueza*. Anthropos-Universidad Autónoma Metropolitana.
Schrödinger, E. (2017). *¿Qué es la vida?* Tusquets.
Shattuck, R. (1991). *La época de los banquetes. Orígenes de la vanguardia en Francia: de 1885 a la Primera Guerra Mundial*. Visor.
Simmel, G. (1986). *El individuo y la libertad*. Península.
Simmel, G. (2017). La trascendencia de la vida. In C. Sánchez Capdequí (Ed.), *La creatividad social: narrativas de un concepto actual* (pp. 213–229). CIS.
Thiebaut, C. (1996). La mal llamada postmodernidad (o las contradanzas de lo moderno). In V. Bozal (Ed.), *Historia de las ideas estéticas y de las teorías artísticas contemporáneas* (Vol. II). Visor.
Watson, P. (2002). *Historia intelectual del siglo XX*. Crítica.
Winnicott, D. W. (2013). *Realidad y juego*. GEDISA.

4

The Originality of Axial Civilizations and the Free Play of History

1 Play Constitutes the Origin of Human Nature, Creativity, Culture, Civilization, and History

Play, an Activity Specific to Mammals and Humans

Playing is an activity that animals and human beings have in common, which does not mean that no deep differences exist between them. The greater attention of parents has made it possible, in mammals and particularly in primates and humans, to intensify the protection given to newborns and imply that they are free from direct selective pressures. Therefore, when animals' basic needs have been met and they do not find themselves in a stressful situation, as a response to that relaxed relationship, play offers them an innovation potential and, consequently, change processes ultimately arise from care, imagination, and learning. In the case of humans, specifically through maternal care, children receive care, love, warmth, and emotional security, essential for the creation of culture, history, religion, morality, science, art, and literature (Maté, 2004, p. 152; Bellah, 2017, pp. 127–138).

J. A. Roche Cárcel, *Creativity and Time: A Sociological Exploration*, Palgrave Studies in Creativity and Culture, https://doi.org/10.1007/978-3-030-84838-5_4

While it is true that play is an activity common to animals and human beings, these also reveal remarkable differences in this regard. Certainly, for young primates, playing is associated—as in humans—with an emotional element of pleasure (Maté, 2004, p. 152). However, in general, for the rest of the animal world, its simple forms shape it as a merely physiological, biological, or physical phenomenon—mainly based on movement. This becomes particularly visible in the presence of exaggerated movements within play sequences, since they are a collection of "translocated, extravagant, fragmented, and clumsy activities." Instead, playing with rules (*games*) is exclusive to humans (Maté, 2004, p. 152), for whom it additionally becomes a meaningful function (Huizinga, 1987, p. 12).

When it comes to children, play largely resembles that of animals, even though it adopts peculiar, more typically human, functions. Firstly, because it releases them from their early indetermination between subjectivity and objectivity, which provides them with an increasingly diversified perspective about the world (Winnicott, 2013, p. 42 ff.; Joas, 2013, pp. 164–167; Bellah, 2017, p. 713). In this sense, young children who play find themselves in a quasi-estranged state, which resembles the concentration experienced by older children and adults. Thus, at the time of play, they are in a place that they can hardly leave and which admits no interferences whatsoever, an area outside the individual which is not an internal psychic reality, though not the outside world either (Winnicott, 2013, pp. 91–92).

Secondly, play has as its purpose to ethically develop the child, since, thanks to play, the child learns to adopt the role of the other (Mead, 1991, p. 155; Bellah, 2017, p. 713). Nor should we forget that, while playing, children are highly focused and own a very intense energy (Bohm & Peat, 2007, p. 61), while simultaneously exciting their body and obtaining pleasure (Winnicott, 2013, pp. 91–92). Furthermore, children's art lies close to play; after all—as pointed out by Freud (1908, pp. 1343–1348)—each child who plays does so as if they were poets creating a world of their own, albeit clearly distinguishing what is play from what is reality. In any case, the young child finds the imaginary universe as important and attractive as the real one (Bellah, 2017, pp. 61 and 139).

Finally, drawing a comparison between primates' playing and that of children shows that the latter play pursuit or fight games; develop a wider variety of motor and rotational skills; investigate and manipulate objects more often; and build toys, play being mums, carry out symbolic and sociodramatic playing activities, and use words with which they wish to reject rule-based play—that is, games—(Maté, 2004).

In a nutshell, it could be said that the difference between play in animals—including primates—and in humans is one of degree. Human ludic activity thus constitutes an imaginary, symbolic, sociodramatic, and even political construction, insofar as it seeks to break the established rules. To this must be added that it is "the play of play," a metagame; note that conscience intervenes and develops therein (Duvignaud, 1982, pp. 72, 84 and 86)—we will explain later that this also becomes essential to understand play in the Axial Age. What happens in this domain also applies to the emotions shared by animals and humans which are transformed by the brain of the latter into a feeling, a sensation, or an emotional state which becomes conscious (Damasio, 2010; Moscone, 2012, p. 57).

Definition and Characteristics of Play

At this point, it seems advisable to specify what we mean by the concept of play and also to clarify what features define it, for which purpose we will establish a conceptual map thereof. The definition by R. Caillois (1986, pp. 37–41) appears to be relevant for our first aim. According to this author, play organizes a free, separate, uncertain, unproductive, regulated, and fictitious activity of four possible types, depending on whether the role of competition (*Agon*), chance (*Alea*), simulation (*Mimicry*), or vertigo (*Ilinx*) prevails in them. D. W. Winnicott (2013, pp. 10 and 81)—following H. Mead (1938, p. 192)—in turn distinguishes between the kind of play which is strictly defined by the rules that govern its operation (*game*) and the one which develops freely (*play*); as well as between *play* (noun) and *playing* (nominalized verb, the act of playing).

Following are some of the key aspects that define play.

It implies the characteristic self-motion of the living. More precisely, it performs a swinging movement that repeats on an ongoing basis, a permanent toing and froing like the play of waves, a movement linked to no purpose or goal whatsoever. This means that, even if play in itself includes the reason characteristic to humans, play rationality is purpose-free. In short, play constitutes a self-representation of the playing movement (Gadamer, 1996, pp. 20 and 67–68). In fact, it implies a free creative activity (Huizinga, 1987, pp. 19–24; Duvignaud, 1980, pp. 7–9; Winnicott, 2013, p. 99). New thoughts therefore arise as a mind game, since it is in the mind that the free movement of conscience and attention becomes closely interrelated with the free play of thinking. After all, creativity needs both types of freedom, which are essentially only one (Bohm & Peat, 2007, pp. 60 and 240). In this connection, it creates an own order with autonomous rules which comes under pressure with freedom, uncertainty, and chance (Huizinga, 1987, pp. 19–24). Thus, although, on the one hand, the rules proposed by the game are meant to be obeyed (Gadamer, 1996, p. 71), on the other hand, they can and must be freely changed (Duvignaud, 1980, pp. 7–9; 1982, pp. 23 and 70).

No wonder play temporarily escapes ordinary life and, for that reason, follows its own trend (Huizinga, 1987, pp. 19–24). In this regard, its time and space differ from those of everyday life, as it is locked within itself and limited (Huizinga, 1987, pp. 19–24; Gadamer, 1996, p. 67). In effect, it occurs within a well-defined space and time, since play episodes begin and finish; what is more, it stages a continuous space–time experience (Winnicott, 2013, pp. 83 and 96; Bellah, 2017, p. 140). As for space, play always leaves a space that needs to be filled—like in *Genesis*, since the fact that things are unnamed enables humans to have the ability for creation—which is why it permanently finds itself in transition (Gadamer, 1996, pp. 20, 21 and 73). And, regarding time, it neither regenerates nor revokes the existing institutions, since it is a useless activity, empty of any content, which opens a hole in everyday life (Duvignaud, 1980, p. 37). It therefore comes as no surprise to check that "0 intentionality" (Duvignaud, 1982, pp. 25 and 126) can only be reached in play, which makes it similar to a party, insofar as, despite being "devoured" or "digested" by organized institutions, it appears in an a-historical moment and as an a-structure within social structures.

Otherwise, play entails an excess (Gadamer, 1996, p. 20) that is intrinsically exciting and, paradoxically, precarious (Winnicott, 2013, p. 98). Not in vain, it generates an extreme intensity of sensations, emotions, feelings, or ideas, along with a powerful mental and intellectual energy (Duvignaud, 1980, pp. 37–39). Neither should it seem strange to us that it results in a communicative act which does not know the distance between the one who plays and the one who simply looks. As an example of this, one of the basic impulses of modern art consists in doing away with the distance that exists between the audience, the consumers or the public, and the work (Gadamer, 1996, pp. 69–70).

Lastly, play builds a space of conflict, but also of conciliation and agreement, between two diverse and plural elements (Gregori, 2002, pp. 43–46). Between these poles, a sign of duality, lie—as shown earlier—rules and freedom, the latter as opposed to chance and uncertainty, the excessive and the precarious, as well as the theoretical limit between the subjective sphere and what is perceived in an objective manner. Likewise, play finds itself at the crossroads between the outer and the inner worlds, in that no man's land where subjective concerns and everyday life meet again, the key required for linking representations to facts and for building the prototype of every significant cultural experience, beginning with art and religion (Winnicott, 2013, p. 96; Duvignaud, 1980, p. 33; De Carolis, 2017, p. 109). We must not forget either that the "play impulse" develops between sensibility and reason, two of the constituent elements of human nature, trying to find a balance between them so that neither predominates (Schiller, 1990, pp. 363–365; Gregori, 2002, p. 44) and without preventing them from expanding freely (Schiller, 1985, pp. 89–90; Gregori, 2002, p. 45).

In connection with the previous discussion, conciliation or conflict arises between two types of play: the impulsive one, which is original; and another which constitutes an implementation of the original impulse. Hence, in play the origin is associated with the conflict, between sensibility and reason, and a relationship is imagined with History, which in turn is understood as temporality—origin—and conflict (Gregori, 2002, p. 47).

Play, the Origin of Creativity and Every Human Activity (Myth, Rite, Religion, and History)

From the perspective presented here, it is important to highlight that play is universal and common to children and grown-ups and, above all, that it represents one of the origins of humanization, of creative activity itself (Winnicott, 2013, p. 100), as well as of culture and civilization.

Of humanization, because play—and the state-of-mind predisposition that it generates—becomes a fundamental task in human life; it is a basic form of life (Winnicott, 2013, p. 96). And this is true to such an extent that, as F. Schiller writes in *Kallias. Letters on the Aesthetical Education of Man*, the latter "only plays when he is a man in the true sense of the word, and he is only a fully-fledged man when he plays." In fact—as stressed by this same philosopher—although animals play too, typically human play is characterized by developing in the space onto which reason and sensibility converge and, by fulfilling that function, it humanizes (Schiller, 1990, pp. 241 and 363–365). In relation to this, it deserves to be highlighted that play is a creative experience which starts from personality, while simultaneously building the person (Winnicott, 2013, pp. 96–100). After all, it arises as a new type of ability with a large potential to develop more faculties (Bellah, 2017, p. 126).

Regarding play understood as the origin of creativity, this activity is indeed the prototype of any creative behavior, even that of the "pecuniary" modern human being—the heir of "barbarian temperament"—that T. Veblen describes in *The Theory of the Leisure Class*, since it represents an ideal type defined by the inclination toward sport, chance games, and the tendency to believe in luck, as well as by the pleasure derived from emulation without an aim (Veblen, 2008, pp. 242–276).

Neither can we understand culture without its playful component (Gadamer, 1996, p. 66), the same holding true for civilization (Caillois, 1986, p. 13). In this respect, play becomes the original source of all human cultural systems, that is, of myth and ritual, of law, of poetry, of wisdom—of art—and of science (Bellah, 2017, p. 122); it thus lies at the origin of culture (Duvignaud, 1980, p. 28; 1982, p. 40).

In effect, language, science, and culture in general equally come from play, or, expressed differently, from the huge creative elaboration and potential that playing brings with it (Bellah, 2017, pp. 162–163). This especially applies to art, linked, in the reflections of I. Kant (2001, pp. 132 and 258), to play, the latter being connected with work as well. According to J. Dewey (Sabariz, 2018), in humans, work and play are not opposites; they actually depend on each other (Bellah, 2017, pp. 119, 692 and 713–714). In other words, play is an autonomous activity—without any external referents (Gadamer, 1996, p. 20 ff.)—and also a disinterested one, but it implies a certain amount of work at the same time. Also Schiller—who follows along Kant's lines in this field—considers that, even though human play starts with its physical part, it may evolve into an aesthetic kind of play in which all the human spiritual and cultural abilities are released (Bellah, 2017, p. 691).

Nonetheless, among all these cultural, civilizing, and historical creative activities what stand out are myth and rite, together with their cultural implications derived from play, to which they owe their capacity to anticipate the future. In turn, these last two activities constitute the source of the instinctive forces of civilized life: law and order, commerce and profit, trades and art, poetry, wisdom, and science (Huizinga, 1987, p. 16; Bellah, 2017, pp. 5 and 163; Duvignaud, 1982, pp. 41 and 72).

Similarly, religion is generated from mammals' play, and more specifically—as we already explained earlier—thanks to parental care. The environment protected by maternal care rids children of selective pressures, which allows them to focus on playing and creating freely. For this reason, the ritual takes place in a relaxed ambiance, which does not exclude the fact that a considerable effort is needed to create such a calm play environment, contrary to animal play, where participants must necessarily be nourished and safeguarded.

As will be explained in the next section, this is precisely what happens in axial civilizations, where the stability and the guaranteed food provided by farming, stockbreeding, handicrafts, and commerce, as well as the irruption of cities, permits the emergence of a series of intellectual and religious elites, a "creative class"—as R. Florida would call it (2010)—strongly centered on creating a universe that is different from the previous one.

2 The Link of Play and Creativity with Axial Civilizations

The Utopias of the Axial Age Come from Play

> Then I—Wisdom—was constantly at his side. I was filled with delight day after day, rejoicing always in his presence. (*Book of Proverbs* of the Old Testament (8:30))

As suggested by R. Bellah, in *Religion in Human Evolution. From Paleolithic to the Axial Age* (2017, pp. 700 and 710), nearly all Axial Age utopias have a significant play dimension. In relation to this, the US philosopher wonders "whether all utopias are nothing but a sort of feigned play in which one can imagine a world that constitutes in itself a relaxed environment where the usual tensions of life become suspended."

It is our aspiration in this chapter to succeed in confirming that link between play and axial civilization, which serves as a starting point, describing the "meaning correspondences" that connect the playing activity itself, creativity, and the Axial Age.

The archaic community plays as animals and children do, that is to say, within a framework of order, with tension and in motion, solemnity, and enthusiasm (Huizinga, 1987, p. 31). The degree of civilization is reached by gradually replacing the primacy of *Ilinx* (vertigo) and *Mimicry* (simulation), which is characteristic of prehistory by the preeminence in social relationships, in the first society-states that do not abandon order, of *Agon* (competition) and *Alea* (chance) (Caillois, 1986, p. 165). In all those societies, disorder was temporarily disarticulated through play, and by submitting the latter to mockery and laughter, by inserting it into narrated or dramatized fiction which causes that effect or, to put it in another way, by substituting real transgression for a fictitious one (Balandier, 2014, p. 112). If Bellah is right, it is in such feigned play that utopia may arise.

And that is exactly what happens in axial societies, in the "time-axis," in the axis of universal history, as K. Jaspers calls them. Indeed, they stage a very profound, historical change in the spiritual process that occurred between the years 800 and 200, more precisely, toward 500 B.C. and until the first century of the Christian era. As a matter of fact, the deepest cut in History takes place during this utopian and highly creative period which relies on the universal. There lies the origin of today's human beings, that is, the fundamental categories with which we still think and live, together with the world religions that are known to us at present (Jaspers, 2017, pp. 17–18; Eisenstadt, 2017, p. 75). More specifically, the six universal religions—whose faithful currently continue to be numerous—are the Confucian, Hinduist, Buddhist, Christian, Muslim, and Jewish religious ethics (Weber, 2017, pp. 344–345).

Well, this Axial Age can be defined in accordance with the characteristics described in the following.

Its origin—the same as that of creativity—is grounded on crisis (Jaspers, 2017, p. 28), and in the extreme situations that it leads to, which is what hastens transformation. For instance, the formidable creativity of the Israelite religion partly arises as a response to the Assyrian challenge, whereas the Indian book *Manu* was probably written during the times of uncertainty and the chaos that followed the overthrow of the Mauryas as an attempt to preserve the privileges of the Brahmans (Bellah, 2017, pp. 383, 461 and 677). Nevertheless, the crisis triggered by that period was subsequently followed by other crises, in such a way that the axial period—just like every other historical stage, after all—ended up, according to R. Bellah, being a "failure," since History continued along its course (Bellah, 2017, pp. 357 and 461), creating conditions other than those of the time-axis which subsequently prevented or hindered creativity or at least minimized its intensity.

The Axial Age establishes an original creative period, insofar as it simultaneously refers us back to an origin and to the originary. It is true, for instance, that Chinese thinkers tried to return to the old, where they found deeply innovative cultural possibilities and, especially, a new oral vocabulary. Furthermore, during the Zhou period, the old was first used to criticize the present, while Taoist primitivism opted for the simplest times of the past in such a way that, with its evocation of childhood's ingenuous spontaneity—a particularly playful and creative phase (as will be seen in the next chapter)—it personified the adolescent crisis of Chinese society more closely than any other school. For its part, Israelite utopia relied on the tough criticism exercised by prophets against the conditions in which the kingdoms of Israel and Judea found themselves (Bellah, 2017, pp. 498, 557, 700). In any case, Israel and Greece are "seedbed societies" (Parsons, 1967, p. 507; Bellah, 2017, p. 721) which have left a deep mark in the future of the current ones.

The Axial Age additionally constitutes an intermediate phase between two periods of great Empires and, consequently, represents an interval of freedom (Jaspers, 2017, p. 85) and sociocultural transition. In this regard, for Taoism, each human being "is the king of their inner country," a moral being advocated which boosts individual freedom and responsibility: "My destiny is in me, not in heaven" (Schipper, 1995, pp. 507, 523 and 526). Instead, Jewish divine creativity was essentially free, as it gave rise to the world through a purely volitive act, without forgetting that it always remains optional, contingent, and free. The great legacy of Greek civilization to the West is the conception about the dignity—*Eleuthería*, freedom—of the human being as such, of any *ánthropos* (Festugiére, 1986).

This stage owns a huge capacity to imagine things other than the existing ones (Bellah, 2017, p. 480), especially utopias; the Taoist literature of the Han period even grants a central value to the imaginary (Schipper, 1995, p. 514). Added to this, a new ideological attitude toward change develops in the Axial Age, accompanied by a constant tension between the creation of change and its rejection or the intolerance thereof, within the same symbolic institutional framework (Eisenstadt, 1986, p. 402). This is what happens, for example, with the luminous creativity of Indian Vedic literature, which features an inexhaustible "ontophanic" nature—showing how the being appears (Eliade, 2011, p. 149). In turn, Taoism sees the universe not as the result of a creator, but as a spontaneous creation from chaos, an evolution in perpetual change "by itself" which flows in parallel to a retrograde force, so that the progress and regression cycles become balanced. The human being occupies an insignificant place in this ever-changing and backward-moving universe (Schipper, 1995, pp. 514 and 521).

The Axial Age differentiates the past, the present, and the future, creating a discontinuity between the key time dimensions and the subsequent need to find a way of building a bridge between them. Nonetheless, although the nature of this bridge, whether it is cyclical, historical, or apocalyptic, turns out to be highly versatile across the different civilizations, what all of them share is precisely the existence of some discontinuity (Eisenstadt, 1986, pp. 388–398). By way of example, Buddhism starts from the present time where one lives and gives priority to present acts (Robert, 1995, pp. 445 and 451), whereas the Jewish religion, guided by History and Time (Eliade, 2011, pp. 44–46), constitutes "the religion of time," since the Jews are the true "builders of time," and its biblical Creation in itself becomes a historical event, the prologue to the rest of History.[1]

The Jewish people additionally create a temporal modality that draws a clear distinction between past, present, and future, without forgetting that the world stops being eternal and unalterable, insofar as it is created, while its current orders represent a construction of human action. In effect, the world is transformed into a historical product, into a means for human action which leaves behind the idle and inactive universe of Paradise. It thus becomes a historical product (Weber, 2017, p. 464) and is consequently determined by uncertainty, insecurity, and freedom, whereas human life moves along linear time.

The Axial Age reaps the fruit of the first inventions but lays the foundations for another type of cognitive innovation, that is, for a more abstract conceptual level (Donald, 2017, p. 70). Not in vain, the concepts of immanence and transcendence introduced are categories of thinking already (Campbell, 2019, p. 79). Furthermore, in the case of Old Testament history, it has an antinatural nuance, since it consists in the confrontation between Yahweh and the cults of Nature—the expulsion from Paradise, the curse of the soil, the defamation of the snake, and Eve's condemnation. Antimaterialism also becomes evident in Plato, who writes in *Timaeus* that the world is made up of pure ideas and impure immaterial substances. Confucianism defines moral as the discovery of the sense of being in the deepest self (Vandermeersch, 1995, p. 570), whereas the Hindu religion portrays an intermediate personal conscience between the chaotic flow of sensorial data and the abstract references of rational metaphysics (Hulin & Kapani, 1995, p. 400). Finally, in

---

[1] The information referring to Jewish religion and Greek Civilization has already been described with the corresponding citations in previous chapters.

Buddhism, its founder exposed a "mental discipline" that he had reached through the development of his intimate abilities. In fact, he is more interested in internal things than in external ones (Robert, 1995, pp. 440 and 445).

The history of the Axial Age is full of attempted fusions—and even universalizations. For instance, according to Taoism, the individual and the world constitute a unity (Schipper, 1995, p. 526), while the Hindu religion establishes no rigid boundaries between the visible and the invisible, the animate and the inanimate, the human being and the animal (or the plant), and between the actual human consciences, as if all things were driven by a single breath. Therefore, for Hindus, the pairs of opposites that constitute the universe merge into an impersonal and neutral power (Hulin & Kapani, 1995, pp. 338 and 398).

Axial Age history also shows frequent examples of tensions, dualizations, and antagonisms, since it constantly opposes change and the intolerance thereof, or orthodoxy, on one side, and division and heterodoxy, on another. Alongside this, the ancient and immemorial is contrasted with the subversive and the heretical (Knöbl, 2017, p. 294; Sánchez Capdequí, 2018, p. 4). It likewise develops a dualized vision of the world and strong antagonistic trends which have played an essential role in the history of mankind ever since (Eisenstadt, 2017, p. 79). This is what happens, for example, in Greece with a clear awareness of polarization, of dualism in the world and in society, which ultimately leads to classifying all beings and things according to a set of oppositions.

However, one of the most important tensions is perhaps that established between the mundane or immanent order and the transcendental one (Eisenstadt, 2017, p. 75), with which the need to reconstruct the mundane order appears as well (Knöbl, 2017, p. 284). That explains why, although the old unity between the political and religious powers continues to exist (Bellah, 2017, p. 340), a profound reorganization of the relationships between the political power and the higher transcendental one is taking place, since the former, which stands for the mundane order, is generally regarded as inferior to the transcendental one and is restructured pursuant to the precepts of the latter (Eisenstadt, 1986, p. 394). This is exactly what happens in the case of the new axial redemption religions, where the tensions with the political orders of the world surely became heightened even to a greater extent because of the fraternity ethics that they advocated (Weber, 2017, p. 417). This applies to Hindus, for instance, since the pairs of opposites that constitute the universe merge into an impersonal and neutral power. This is so because the Brahman owns an absolute transcendent dimension, situated beyond time, while immanence lies in the heart of the created world. This accounts for the establishment of a cut between the Supreme Lord himself, creator of the universe and the keeper of its immanent order, the one who produces it and reabsorbs it periodically, and the *Mandalins*, that is, the "cosmic regents" who belong, like human beings, to the world and to time (Hulin & Kapani, 1995, pp. 339–340 and 405).

On the other hand, the relationship with the transcendent god and with the world, understood as a corrupt and ethically irrational creature, results in the loss of sanctity of tradition and the need to work on a permanent basis, always in an incomplete and unfinished manner. This is well exemplified with the Hellenic unstable conception of the world and the human being, according to which, from the standpoint of C. Castoriadis, the "social being is chaos, indetermination, and creation" (Cristiano, 2009, p. 105); with a creation—Hesiod's Theogony—which leads to a life eternally renewed through creativity; or with a Constitution, always in process and never closed or fixed. Something similar can be found in the Islamic religion, which moves "from creation as proof of the Creator to creation as proof of the creature." After all, creation is carried out by God and only by him; and then, God chooses the human being to act as his vicar on earth, which is why the latter has to make decisions on his own, time after time (Guellouz, 1995, pp. 257, 261 and 287). According to the Jewish religion, divine creation, a model of the human one, is fun for God, since he sees it as a big game thanks to which all that exists takes shape and through which all of this becomes a show. In this connection, creation evidences arbitrary laughter on the part of God; instead, for humans, it reveals the domain of necessity and—as seen earlier—of freedom, while the risks arising from the creation game represent the figures of chance (Balandier, 2014, p. 23). Therefore, the peculiarity of human beings, what builds them as such, is their ability to imitate God and his way of exercising the power of creation (Boorstin, 2008, p. 71), to ethically and rationally beat and dominate everything that exists around them.

That is how we plant the seeds of what will germinate in the modern age as the rational objectivity of progress (Weber, 2017, p. 397).

The Creative Agents of the Axial Age

This germ of rationality established by the Axial Age was created, fixed, and articulated by means of a relatively new element which arose within the bureaucratic structures that characterized the preagricultural and agricultural societies of the Axial Age (Donald, 2017, p. 47 ff.). That is to say, through a new type of intellectual and clerical elite that comprises Jewish priests and prophets, Greek philosophers and sophists, Chinese literati, Hindu Brahmans, Buddhist Sangha monks, and Muslim mullahs (Eisenstadt, 1986, p. 389).

These creators of the axial period had a network of continuators—of re-creators—who expanded the so-called universal religions generated during the Axial Age. Among them stands out the bureaucrat organizer of the world in Confucianism, the wizard organizer of the world in Hinduism, the mendicant monk who travels around the world in Buddhism, the warrior conqueror of the world in Islamism, the traveling salesman in Judaism, and the wandering craftsman in Christianity. They are actually the ideological keepers of an ethics or a doctrine of redemption that very suitably matches their social position (Weber, 1997, p. 190). Well, from my perspective, they could be divided into two large groups: the bureaucrats and the itinerants or nomads, the institutionalizers, and the carriers or communicators of ideas.

In any case, the true creators of axial time isolated themselves; they critically and reflexively questioned reality, and also observed and conceived a vision that went beyond everydayness, as Buddha did, for instance, when he criticized the caste system and certain Brahmanic beliefs (Robert, 1995, p. 442). Simultaneously, in tune with this transcendental cosmovision, they became aware of the need to actively build the world, which led them to promote protest movements, to drive conflicts, and to encourage changes which made it possible to create new utopian models of social and cultural order (Eisenstadt, 1986, pp. 385–389 and 400 and ff.).

At a mental, social, and cultural level, these elites managed to move from a magical world that was mainly agricultural and dominated by sorcerers to another religious one of an urban nature where priests played a relevant role. And, more importantly, in particular for the history of civilization, they succeeded in partially replacing magic by reason, or, better still, they added to the understanding and manipulation of this world and that of the afterlife the rational path alongside the traditional magical one. A significant example in this respect can be found in the Greek logos created by Western-style philosophers that is inherent to the creativity process. This logos—the same as creativity—expresses a way of thinking, of telling or calculating, of relating and being.

However, next to this rationality, powerful irrational forces of an external or internal origin also existed in Greece which destroyed humans to such an extent that their actions—including the creative ones—may be not only rational but also irrational, wild, and with unforeseeable effects.

Consequently, according to M. Weber, magic and reason are going to coexist from axial civilizations, insofar as, although the former has a rational purpose, this is sought by irrational means, while religion is characterized by objectives which gradually become more irrational and are pursued by growingly irrational means. Anyhow, in its original form, the action derived from magical or religious motives will from now on be oriented toward this world (Weber, 2017, p. 63).

Nonetheless, the emergence of priests as a social element other than sorcerers clearly distinguishes the new axial world from the old one. After all, the name of priests is used to refer to the staff members of a permanent enterprise organized on a regular basis with the aim of influencing gods, which opposes them to sorcerers, who exclusively own isolated and circumstantial resources. In turn, the prophet proclaims the truth of salvation by virtue of the legitimacy associated with his personal revelation (Weber, 2017, p. 89 and 114).

The novelty also has to do with the spatial context where this transformation takes place, since peasants are very unlikely to foster the creation of a non-magical religiousness. On the contrary, it is the petty bourgeois—based on his economic lifestyle—who tends to generate ethical–rational religiousness, and this should come as no surprise, since the life of a petty bourgeois, at least that of an urban craftsman and a retailer, no longer depends on nature, unlike what happens with peasants' life. This explains why petty bourgeois do not see the subordination of magical influence on irrational spirits in the same way as peasants do (Weber, 2017, pp. 142–157).

The Originality of the Axial Age

After describing the key aspects of play, creativity, and axial civilizations, the analysis of their interrelation(s) has made it possible to detect the "meaning correspondences" (in M. Weber's words) between them, and also to understand the Axial Age as a very original period able to build a time narrative, as will be explained here in greater detail.

The Axial Age is indeed deeply original—this being a fundamental concept within creativity—as verified here, since it returns to an origin and constructs something originary, focused on the future. Thus, as in creativity, which relies on crisis and on its extreme situations, crises and the new conceptions introduced by them also drive the religions of the axial period which break the normative frameworks imposed by their society. By way of example, the originality of axial civilizations becomes visible in the return toward the origin of Chinese civilization, for the purpose of looking critically at the present, and in the originating nature, as a seedbed, of Israelite and Greek civilizations.

This originality of the Axial Age additionally comes as a result of its creative freedom, which brings it close to play and creativity—naturally free too—without forgetting that, since play is a permanently transitional activity organized in a space that needs to be filled, the axial civilization represents a sociocultural transition situated between two normative and authoritatively instituting empires. For this reason, it represents a stage of profound freedom in the time-axis, as shown by the conception of dignity—*Eleuthería*, freedom (of the human being), the great contribution made by Ancient Greece to Western civilization, or by Taoism, for which each human being "is the king of their inner country" and, therefore, it advocates a moral which boosts freedom. This freedom otherwise entails that humans become makers of an unfinished creativity, as it takes place, for instance, in the Jewish religion, established from the optional, the contingent, the uncertain, and the free. Or, as it happens in Ancient Greece, which conceives an essentially unstable human world where life must be eternally renewed by means of creativity and which coherently ends up elaborating an incomplete political constitution in need of constant updating.

Together with this concept, a correspondence exists between the specific order created by play and the ability of axial civilizations to imagine different things—a central value is even assigned to the imaginary in the Taoist literature of the Han period. In turn, Jewish antinaturalism detaches itself from nature, and so does Platonic idealistic antimaterialism, which shapes a world made of pure ideas and impure matter. Finally, axial religions take advantage of transcendence to generate another kind of universe located beyond everydayness—the same applies to play—and reality, which is observed differently, not in a profane way, but in a sacred one.

The Axial Age additionally shows a new ideological attitude toward change itself as it occurs, for example, in the luminous creativity of Indian Vedic literature, which features an inexhaustible "ontophanic" nature (showing how the being appears), or in Taoism, which sees the universe not as the result of a creator, but as a spontaneous creation from chaos, that is, an evolution in perpetual change "by itself."

This imaginary capacity—like play—is fueled by the huge emotional, mental, and intellectual energy and by the powerful emotional effervescence that the Axial Age has deployed. Hence the time-axis can be understood as an extraordinary stage in the history of mankind, so much so that we still continue to be fed by it.

The Axial Age Organizes a Time Narrative

The Axial Age actually builds a time narrative which has connected the present with the past and the future, thus merging the origin and the originary, which relates to its conception of space and time. In fact, the time narrative of the Axial Age occurs—and a meaning correspondence also exists with play in this respect—within a limited, locked, and continuous space–time lying outside everydayness, where the creative intellectual elites voluntarily isolate themselves from their society so that, from such distance, they can orient themselves toward the world, actively build it, and reinforce it. This is what the Chinese Civilization does when it returns to the past seeking to criticize the present from that past.

To this must be added that axial religions establish a historical bridge which constitutes in itself a time narrative, even if the latter has a diverse cyclical, historical, or apocalyptic nature. Thus, the time narrative of the Axial Age materializes with the sort of swinging movement that—the same as play—performs and goes from the past to the future visiting the present on the way, and from the crisis which triggered it to the one that closes it. The first idea is what can be attested, for instance, in the Chinese civilization, which goes back to childhood—an outstandingly creative stage—looking for simplicity and fleeing from the complexity of its period. The second one may perhaps lead to think—as R. Bellah does—from the perspective of its time, that the axial civilization was a failure, more or less in the same way as play eventually becomes a useless activity which neither regenerates the past nor revokes the existing institutions. However, insofar as myth, a product of play, manages to anticipate the future, the myth of axial religions likewise succeeds in implementing a prophetic sense, and even in creating the future. It seems advisable to remember in this regard that its utopias stem from the feigned, imaginary, and creative play where the transgression of the real is replaced by a fictitious one; they deal—like all of them—with the future (Hudson, 1982, p. 167; Ricoeur, 1988, pp. 119–120). And thus, everyday reality follows its course, albeit no longer from the same form as before, but contaminated—or, rather, enriched—by the desire for utopia.

In addition to this, the typical discontinuity of time which becomes evident, for example, in Buddhism, starts from the present where one lives and gives priority to the acts which take place within that temporal modality. It is also reinvented in the Jewish religion, guided by History and Time, since it conceives linear time, which sees the world as a historical construction and, accordingly, as a means for human action, and which imagines a creation of the universe that turns into the prolog of History.

It should consequently come as no surprise that this time narrative—just like play—has a dialectical nature, since it merges and simultaneously tenses, confronts, dualizes, and antagonizes. The first and foremost thing happens because the Axial Age is a universalization stage. After all, similar to the way in which play is universal, because it is implemented by men and women, children and grown-ups, universalization underpins the Axial Age as well. Therefore, the creativity of this stage has originated in different parts of the planet, and in a variety of civilizations, exactly as it happened when agriculture arose.

Deep thinking about unity emerges in two of those civilizations: China, the birthplace of Taoism, where fusion reaches its maximum intensity, because the individual and the world form a unity; and India, where pairs of opposites that constitute the universe merge in an impersonal and neutral power.

On the other hand, it deserves to be remembered that, apart from associating origin and conflict, play develops in a no-man's land where the opposite poles become inseparably intersected. Thus, the time-axis contrasts but at the same time links order and freedom, uncertainty and chance, the inside and the outside, the subject and the object, fun and work, the mundane order and the transcendent one, space and time, the past and the present, the present and the future, reality and utopia.

In this connection, within this last polarization, in the same way as play is driven by a purposeless rationality and by its link with effort, this ludic rationality gives rise to the ethical rationality of axial religions, which will in turn become the seeds of modern work rationality—Weber dixit.

Furthermore, just like play can be described as a communicative activity which removes the distance between players and observers, something similar happens in axial religions with such social types as communicators and carriers who, as described by M. Weber, act as continuators of the pristine creators.

The second notion, that is, the fracture or dualization of the world, becomes expressed in the Axial Age through the tension between transcendence and immanence. This is present, for example, in the Islamic religion, according to which creation was performed by God and only by him, whereas God chooses the human being to act as his vicar on earth, which is why the human being has to make decisions on his own, time after time. In turn, the Indian religion portrays the Brahman as having an absolute transcendent dimension situated beyond time, insofar as immanence lies in the heart of the created world.

3 Three Foundations of the Axial Age Which Help Understand Our Time

This chapter has served to unveil the "meaning correspondences" existing between creativity, play, and axial civilizations. More precisely, three foundations of the Axial Age have come to light and make it easier to understand our time: it is a very original and creative stage, it organizes a time narrative and intensifies the awareness of what we are, where we come from, and where we want to go.

1

Firstly, the Axial Age stands out as an extremely original—and consequently creative—period, since it returns to the origin and builds something original, focused on the future. Its origin is the period of crisis and transition during which new powerful ideas strongly arise, while its original or originating nature leads us to reconsider the current critical times, for they could be solved with creativity—as suggested by H. Joas. However, what the Axial Age teaches is that this will only be possible if we deploy such creativity freely and bearing in mind that thinking, imagining, dreaming, and creating are free exercises; after all, as Taoism defends, each human being "is the king of their inner country." Furthermore, the Axial Age, particularly the Han period's Taoism, assigned—as children do—a central role to the imaginary and precisely encouraged imagining different things. Among these is a new ideological attitude toward change itself, something which appears as one of the key characteristics in our liquid, evolving, and ever-flowing societies. Neither should we forget that the Axial Age meant a strong and effervescent emotional display, which is why it may be perceived as an extraordinary stage in the history of mankind, so much so that we still continue to be nourished by it.

In any case, attention should be drawn to the fact that creativity—as highlighted by C. Castoriadis—is a permanently unfinished activity; hence the assumption that the exit of the contemporary crisis will be, as it has always been, transitory, that a boundless creativity must be implemented, and that its outcome can never be definitive or complete. In that sense, what happens to creativity resembles the vicissitudes of our representative democracies, always under construction, never fully accomplished—neither in Ancient Greece nor today—and constantly in transit toward an improved and perfected state.

2

Secondly, the Axial Age—like creativity—builds a time narrative, insofar as it returns to the origin, critically detaches itself from the present, and develops utopias which open its societies toward the future. The origin and the originary have indeed merged in it, but this has happened outside the everydayness, since the intellectual elites have separated themselves from their society in order to be able to judge it suitably and to propose new alternatives. Moreover, everyday reality followed its course in the Axial Age, though not in the same way as before, but enriched by the desire for utopia. This might remind us contemporary people that we more than ever need utopias able to transform a social reality for which not many ways out generally seem to exist.

On the other hand, like in the Axial Age, a temporal discontinuity still exists today that, like in Buddhism, prioritizes the acts of the present, which has currently become a tyrant (Beriain, 2008). And we continue to be oriented, exactly as in the Jewish religion, by History and Time, except that the latter has been hugely accelerated and the former has darkened. Some people even think that we find ourselves at the Apocalypse of History, in its final stretch, at the end of the times, thus completing the great book of History inaugurated by the Jewish people, according to whom their stage was the prolog to that great book. My approach, in tune with that of Negri, is pessimistic regarding reason and optimistic in relation to will; I am reluctant to think in this way and see creativity—with its undeniable limitations—as offering possibilities to escape from our chaotic and undetermined situation.

It is worth highlighting in this respect that History occasionally takes leaps and prances around—as animals do when they play—and that its movements are sometimes excessive, just like those of primates. Nevertheless, History—similarly to creativity—constitutes a time narrative which persistently looks toward the past in order to build a present and, from it, the future. One of those leaps taken by the Axial Age was toward the territory and the time of transcendence and, when doing so, it showed a certain distance with respect to this world, thanks to which it has been possible to watch it differently—more critically—ever since. Thus, we have had the chance to question the social organization form, an alternative universe having been imagined in its stead. The leap to which we refer is consequently toward the future, modeled in its origin in the Axial Age. Contemporary society, the original product of that age, is the heiress of that leap and, therefore, has in turn the responsibility to act as the origin that will originate the future.

Two final reflections need to be made in this connection. The Axial Age represents a universalization stage and, as such, has become a germ of the enlightened cosmopolitan ideals and of present-day globalization. One of the great challenges that it faces is the COVID-19 pandemic, which has worsened a crisis that was already very deep, and which accordingly urges us to display a huge amount of creativity and energy if we want to overcome it. Furthermore, if play and the Axial Age can be described as a communicative doing which removes the distance between the player and the observer, doubts arise about whether today's representative democracies can really be transformed partly or completely into participatory democracies, as a segment of society has been insistently demanding.

3

The Axial Age has established that immanence and transcendence are categories of thinking, a mental game, "the play of play"—as J. Duvignaud would express it—a metaplay where conscience intervenes and develops. According to the Hindu religion, this conscience focuses to a greater extent on internal things than on external ones, and it lies inside the person, in a space situated halfway between the chaotic flow of sensorial data and the abstract references of rational metaphysics. Confucianism, in turn, defines morality as the discovery of the sense of being in the deepest self, exactly as Taoism stresses that each individual is the true king of their inner country. On the contrary, the Jewish religion represents one people's awareness, as it occurs in Hinduism or Shintoism, religions of birth, of kinship (Campbell, 2019, pp. 105 y 110).

The Axial Age consequently identified social or individual conscience as an essential value of creativity, an idea that was subsequently taken up by Kant (2001) when he placed creativity—and the aesthetics emanating thereof—in free subjectivity (Duvignaud, 1982, p. 60 ff). Both creativities—social and individual—go hand in hand, though, since individual awareness precisely becomes the foundation of Reflexive Society (Lamo de Espinosa, 1990), of the society that reflects on itself, which detaches itself to make self-criticism possible and, ultimately, to develop alternatives that can leave its lethargies or paralyses behind. The reflexive society is one which plays creatively, starting from its origin, from a revised past, with the aim of questioning the present and building the future.

And it undoubtedly does so with rationality, the root of which lies in the Axial Age. Thus, the latter links the purposeless rationality deployed in play, that is, ludic rationality, to effort. This results in the emergence of the ethical rationality that prevails in axial religions, which will in turn become the genesis of modern work rationality. In any case, the Axial Age has also shown us that the force of rationality does not suffice in creative play, that creativity is set in motion by an intense emotional energy that becomes effervescent, perhaps because of what Simmel used to say: "life calls for more life."

It can therefore be stated that History arises in day-to-day activity, in the folds of everydayness, but also in the deep valleys and the high mountains, in those milestones, in those extraordinary moments—like those belonging to the Axial Age—where an effervescent or, as Durkheim calls it, "revolutionary," energy is deployed.

In this sense, and connected to the teaching of the Axial Age, one might wonder if History is in fact a leap forward, a way to overcome crises and a utopian aspiration to construct what is still non-existing. Hence, despite the seeming fiasco—in R. Bellah's opinion—that the Axial Age entails, thanks to its originality, and by means of its typical time narrative, it has renewed History, our History, and through it, also the actual existence of individuals, societies, cultures, and civilizations.

And yet, there is something else, since it can be inferred from all that was discussed that, on some occasions—the most ludic, original, and creative ones—an extraordinary free play of history takes place. That is exactly what happened in the period known as time-axis, and nothing gives grounds to believe that it cannot occur once again.

Bibliography

Balandier, G. (2014). *El desorden. La teoría del caos y las ciencias sociales*. GEDISA.

Bellah, R. N. (2017). *La religión en la evolución humana. Del Paleolítico a la era axial*. CIS.

Beriain, J. (2008). *Aceleración y tiranía del presente. La metamorfosis en las estructuras temporales de la modernidad*. Anthropos.

Bohm, D., & Peat, F. D. (2007). *Ciencia, orden y creatividad. Las raíces creativas de la ciencia y la vida*. Kairós.

Boorstin, D. J. (2008). *Los creadores*. Crítica.

Caillois, R. (1986). *Los juegos y los hombres. La máscara y el vértigo*. FCE.

Campbell, J. (2019). *Tú eres eso. Las metáforas religiosas y su interpretación*. Atalanta.

Cristiano, J. L. (2009). Imaginario instituyente y teoría de la sociedad. *RES. Revista Española de Sociología, 11*, 101–120.

Damasio, A. (2010). *Y el cerebro creó al hombre: ¿cómo pudo el cerebro generar emociones, sentimientos, ideas y el yo?* Destino.

De Carolis, M. (2017). *La paradoja antropológica. Nichos, micromundos, disociación psíquica*. Editorial Quadrata/Iluminuras.

Donald, M. (2017). Una aproximación evolutiva a la cultura. Consecuencias del estudio de la época axial. In C. Sánchez Capdequí (Ed.), *La creatividad social: narrativas de un concepto actual* (pp. 47–73). CIS.

Duvignaud, J. (1980). *Le jeu du jeu*. Bolland, Paris. Mesa redonda. Nuevos horizontes. El juego. p. 118. Retrieved March 13, 2020 from www.libertario.arte.

bo/biblioteca/Mesa%20redonda%20Nuevos%20Horizontes%20-%20Jean%20Duvignaud%2C%20el%20juego%20del%20juego.html
Duvignaud, J. (1982). *El juego del juego*. FCE.
Eisenstadt, S. N. (1986). *The origins and diversity of axial age civilizations*. State University.
Eisenstadt, S. N. (2017). La dimensión civilizacional en el análisis sociológico. In C. Sánchez Capdequí (Ed.), *La creatividad social: narrativas de un concepto actual* (pp. 73–95). CIS.
Eliade, M. (2011). *Ocultismo, brujería y modas culturales*. Paidós.
Festugiére, A. (1986). *La esencia de la tragedia griega*. Ariel.
Florida, R. (2010). *La Clase Creativa. La transformación de la cultura del trabajo y el ocio en el siglo XXI*. Paidós.
Freud, S. (1908). El poeta y los sueños diurnos. *Obras completas, 4*, 1343–1348.
Gadamer, H.-G. (1996). *La actualidad de lo bello*. Paidós.
Gregori, E. (2002). Individuo y naturaleza en Schiller. La noción de juego. *Materia, 2, Naturaleses*, 35–55.
Guellouz, A. (1995). El islam. In J. Delumeau (Ed.), *El hecho religioso. Enciclopedia de las grandes religiones* (pp. 253–331). Alianza.
Hudson, W. (1982). *The Marxist philosophy of Ernst Bloch*. Springer.
Huizinga, J. (1987). *Homo Ludens*. Alianza.
Hulin, M., & Kapani, L. (1995). El hinduísmo. In J. Delumeau (Ed.), *El hecho religioso. Enciclopedia de las grandes religiones* (pp. 333–402). Alianza.
Jaspers, K. (2017). *Origen y meta de la historia*. Acantilado.
Joas, H. (2013). *La creatividad de la acción*. CIS.
Kant, I. (2001). *Crítica del juicio*. Austral.
Knöbl, W. (2017). El debate de 'las modernidades múltiples, su fundamento sociológico-religioso y sus escollos. In C. Sánchez Capdequí (Ed.), *La creatividad social: narrativas de un concepto actual* (pp. 263–315). CIS.
Lamo de Espinosa, E. (1990). *La sociedad reflexiva*. Siglo XXI.
Maté, C. (2004). El juego en los antropoides y los niños. *Estudios de Psicología, 25*(2), 149–168.
Mead, G. H. (1938). *The philosophy of the act*. The University of Chicago Press.
Mead, G. H. (1991). *Espíritu, Persona y Sociedad. Persona y Sociedad desde el punto de vista del conductismo social*. Paidós.
Moscone, R. O. (2012). El miedo y sus metamorfosis. *Psicoanálisis, XXIV*(1), 53–78.
Parsons, T. (1967). Evolutionary universals in society. In *Sociological theory and modern society*. The Free Press.

Ricoeur, P. (1988). *Time and narrative* (Vol. III, K. Blarney & D. Pellauer, Trans.). Chicago University Press.
Robert, J.-N. (1995). El budismo. In J. Delumeau (Ed.), *El hecho religioso. Enciclopedia de las grandes religiones* (pp. 427–474). Alianza.
Sabariz, C. R. (2018). Una salida al dualismo: Juego y trabajo en John Dewey. *Éndoxa, 41*, 135–156.
Sánchez Capdequí, C. (2018). Las sociologías post 68 y la creatividad social. *Arbor, 194*(787), a435.
Schiller, F. (1985). *Sobre poesía ingenua y poesía sentimental y una polémica Kant, Schiller, Goethe, Hegel*. Icaria.
Schiller, F. (1990). *Kallias; Cartas sobre la educación estética del hombre*. Anthropos.
Schipper, K. (1995). El taoísmo. In J. Delumeau (Ed.), *El hecho religioso. Enciclopedia de las grandes religiones* (pp. 509–548). Alianza.
Vandermeersch, L. (1995). El confucionismo. In J. Delumeau (Ed.), *El hecho religioso. Enciclopedia de las grandes religiones* (pp. 333–402). Alianza.
Veblen, T. (2008). *Teoría de la clase ociosa*. Alianza.
Weber, M. (1997). *Sociología de la religión*. Fondo de Cultura Económica.
Weber, M. (2017). *Sociología de la religión*. AKAL.
Winnicott, D. W. (2013). *Realidad y juego*. GEDISA.

5

The Ages of Creativity

1 From Conflict to Generational Coexistence. Effects on Creativity

Youth and the Conflict Between Generations as a Driver of Creativity

Creative Youth and the Conflict Between Generations

Youth—unlike old age, to which a conservative character is assigned—is the most "tempestuous" and innovative period. After all, even though young people's first impressions and life experiences guide—as during childhood—any of their later experiences, they also reject the acquired ones (Mannheim, 1993, pp. 196 and 216). It comes as no surprise that immaturity is inherent to adolescents and that they attempt to "kill" what has been inherited from their society, all in order to reach the adult category (Winnicott, 2013, p. 215 ff.). Hence, youngsters are not responsible for the world that they have received and that is why they shape "the photograph of the changing times" (Bauman & Leoncini, 2018, p. 17), since they have the most encouraging factors of creative thinking, the

J. A. Roche Cárcel, *Creativity and Time: A Sociological Exploration*, Palgrave Studies in Creativity and Culture, https://doi.org/10.1007/978-3-030-84838-5_5

newest and freshest feelings, and ideas for a new life; society actually needs to be convulsed by their aspirations in order to become renovated and revitalized, to achieve its rejuvenation. However, it is also true that society demands from young people—for them to become fully fledged members of it—that a cure be obtained for such congenital immaturity, which is only possible through the passage of time, with the maturity that the latter can provide (Winnicott, 2013, pp. 215–227), obtained by means of the clash, and the responses that it entails, between youngsters' worldview and that of society itself.

Consequently, the experiences collected by young people throughout their existence are not only accumulated by addition, but also become coupled dialectically and, very often, on a combat basis. More precisely, this fight takes place from those overlapping sediments that, when turned into conscious ones, are memorized or transformed, although the latter only at their most superficial level, since the deepest ones cannot be destabilized (Mannheim, 1993, pp. 216–219). This happens because individuals, during their socialization process, need to rely on the previously existing knowledge and extend it so that such knowledge can become the fruit of mankind's development, that is, of its long learning process. Note in this respect that the main subject of this wisdom is not the individual, but human groups, and, more specifically, the chain of generations (Elias, 2013, pp. 15 and 53).

Let us see then what the concept "generations" means, what role is assumed by youngsters in them, and how those generations work and become linked to creativity (Ortega y Gasset, 1983, p. 149; Ortega y Gasset, 2008, pp. 43–149; Martin, 2008, pp. 98 and ff.; Mannheim, 1993, pp. 196–232; Halbwachs, 2011, p. 112).[1]

A generation—a group of human beings whose age is the same—constitutes a dynamic engagement between the mass and the individual. It additionally represents, together with natural values (as it ultimately relies on the biological rhythm of birth and death), the most important notion within historical evolution, the key which permits to know its movements, changes, and resistances. Therefore, what matters is not that generations succeed one another, but that they coexist and are

[1] This part is an elaboration of mine based on the readings mentioned.

contemporary, and therefore not coeval, which means that they "overlap or combine" and enter into "controversy." This consequently explains why conflict and collision arise as the foundation of historical matters when it comes to any social coexistence and to the dynamic dramatism of societies. Nevertheless, at the base of this observation lies the continuous nature of generational change, alongside the fact that human beings are eternal misfits permanently seeking to modify what they have inherited in order to adapt it to their own sensibility and to endow it with their own singular imprint. Each generation consequently works along two simultaneous dimensions: receiving from the previous generation what they lived through, that is to say, the ideas, values, and institutions, and allowing their own spontaneity and worldview to emerge and develop. Two different types of generations thus exist: the ones who coexist with their elders and those who fight against it. Certainly, in the former, young people declare their solidarity with the old ones and become subordinated to them, both in politics and in science or in the arts, dominated by adult people's experience; hence these times are aged—an issue which I will insist upon later. Instead, the ages—the first modernity—where youth prevails are initiatory and rely on a constructive or destructive belligerence and, therefore, on creativity.

In addition, at the root of the clash between generations stands the existence of three stages in life: the child, the youngster, and the mature or old person. This entails that every present comprises three different temporalities, each one with its own conception, albeit intertwined with one another in an essential hostility: that of 20-year-olds, that of 45-year-olds, and that of 60-year-olds. The present therefore constitutes a unit within a historical time of three different ages, of three lifestyles, each one of which stands for a peculiar way of approaching life. To this must be added that the historical reality is guided by human beings immersed in two different stages of existence which cover about 15 years, a period of time in which the "historical tonality" varies. After the first 30 years of life, a training period in which youngsters build their own biography, between the ages of 30 and 45, adults have already become aware of the world they live in. After all, at this age of gestation or creation and controversy, grown-ups find their own solutions, in contrast with the social universe where they live, and invent not only new ideas but also original social and cultural forms. This innovated, instituent universe will in turn be transformed in the near future into an established one that will restart a new life cycle—the one going from 45 to 60 years of age, a time when human beings (the same as children) leave public life and lose interest in contemporary events—of predominance and command in which citizens preserve what they have produced, by defending it, directing it, and governing it with conviction. Nevertheless, in contrast to this universe which comes into force now, 30-year-old youngsters react against it, thus starting the generational conflict once again.

2 The Rejuvenation of Science and Art

The Rejuvenation of Modern Science

This conflict between generations becomes visually explicit during the first modernity—in the period going basically from 1890 to 1930 (Bell, 1992, p. 117)—when the ties between culture and youth and between the latter and modernity themselves turn out to be more evident (Hobsbawm, 2013, pp. 112–113). However, it extends until 1973, when the second or late modernity starts, albeit differently. Therefore, modernity is young, as it enthusiastically announces unexpected boundaries which are opened, new stories to tell, and, consequently, innovative imagined endings. Thus, novelty—originality—becomes the characteristic feature of the time until World War I (Watson, 2002, p. 19). It additionally arises as a period of "hope" and enormous creativity, since a stage of extraordinary economic growth—known as "golden age"—begins after 1945 (Hobsbawm, 2001, p. 29 ff), and because activities such as relativistic and quantum science, avant-garde art and fashion flourish in interwar years.

These last activities especially developed during the period between 1905 and 1908—"the cubist decade," the "belle époque," "the banquet years" (Barzun, 2001, p. 1005)—that witnessed a renewed cultural strength which had nothing to do with the late-nineteenth-century decadent, negationist, and negative attitude toward the world, seen as distant

(Barzun, 2001, p. 949). More precisely, this stage generated four forms of energy: a practice that reveres violence; the great artistic achievements; scientific innovations; and social reforms—for instance, the current conceptions of the State and of representative democracy.[2]

As for science, two major innovative scientific events can be summarized, the two fundamental pillars which underpin modern physics—the special and general theory of relativity and the quantum theory—regarded as the two most important achievements in the last 100 years (Sánchez Ron, 2000, pp. 49–117; Barzun, 2001, p. 1106; Greene, 2006, pp. 15–180 and 333–335; Chown, 2007, preface, pp. 19–141). The first theory of impact was developed by Einstein in two stages, the initial one known as "the special theory of relativity," of 1905, a revolutionary and innovative proposal about the image of space and time which, from then on, stopped being absolute to become relative. In fact, while the former contracts, the latter expands and, therefore, becomes elastic, which means that the observers who find themselves in a relative movement with respect to one another will obtain different perceptions—all of them true—about distance and time. More specifically, Einstein provides as an example of the new space–time configuration that, for someone who walks past us at great speed, time slows down, that is, it elapses more slowly for that person; and space shrinks, without forgetting that a portion of that space looks like time to us and a part of that time looks like space to us. Consequently, these two dimensions, which formerly seemed to be separated and absolute, now become interchangeable—shaping the two sides of a single coin—and relative.

This is what happens with the speed of light, whose enthronement as the ultimate criterion by Einstein meant that, since then, no physical reality—except for it—was going to be considered absolute. Indeed, the universe that we watch through a telescope, despite extending along the space–time continuum, is actually light, since what stays constant in it is neither space nor the passing of time, but the invariable speed of light. This conception is the one which has replaced the traditional vision of space and time as rigid and objective structures by another new one determined by the relative movement existing between the observer and

[2] I have added science to the three forms of energy identified by Barzun (2001, p. 1005).

the observed. Furthermore, it has also revealed that light does not age and that the passing of time does not exist in the speed of light, which means that the latter constitutes the "foundation," the "rock" on which the universe is supported, on top of which it is built, and that time and space accordingly assume the role of "moving sands."

In 1915, Einstein created the general theory of relativity, a new gravitational conception which replaced Newton's classical universal gravitation theory. According to it, space stops being immutable and alien to its energy-material content to become a curved four-dimensional space–time. More precisely, Einstein specifies that masses—stars like the Sun—because of their weight, curve, or "warp," the space–time existing around them, whereas other masses, for example, the planets or the Earth, freely slide along the space–time driven by their own inertia. Otherwise, this theory, hardly applicable to real situations or to everyday life, is circular, insofar as matter tells the space–time how it has to bend; later on, the warped space–time suggests to matter how to move; and finally, the latter, which has just changed its position, tells the space–time how it must change its curvature. Nonetheless, the total size of the spatial universe in fact no longer remains invariable as it had hitherto been believed. Instead it changes with time, that is, its structure becomes enlarged or reduced.

The second of these scientific innovations, and the one which has most altered the course both of science and of society itself by being omnipresent in everydayness, is quantum physics. The latter arises from the struggle to reconcile light and matter, which is why it essentially contradicts all the previous science. By way of example, the Copenhagen-born Niels Bohr divided the universe into two domains governed by different laws in 1920: on the one hand are the very small things—the atomic world—governed by quantum mechanics; and, on the other hand, the very big things—the everyday world—ruled by the normal or classical laws. In turn, Heisenberg's uncertainty principle unveiled in 1927 that energy and momentum are uncertain; or, expressed differently, we "can no longer know, for reasons of principle, the present in all its details." Furthermore, by applying the uncertainty principle to quantum mechanics we discover that, at a microscopic level, the universe is a "swarming," "frantic," and "chaotic" environment. Therefore, physics has ever since had to address the formal description of the connections between perceptions, in such a way that the physical world loses some of its solidity and fades—vanishes—into the foundations of its structure.

The derivations from the general theory of relativity and from quantum physics likewise made it possible for Hubble to demonstrate in 1928 that the Universe is expanding, which means that, in the past (13.7 billion years ago), there was a time when the whole matter would have been concentrated in a small extension, the *Big Bang* (as the astrophysicist Fred Hoyle subsequently called it), a cosmic explosion that launched to the outside the material contents of the universe (a theory which is now regarded as somewhat mistaken because there must have been no space around the *Big Bang* at the time), the "debris" of that explosion, in the form of many billion galaxies, which continue to circulate outwardly and expand the universe. The end of this journey would possibly be a big cosmic crunch, a great implosion, a universe crammed in a compressed cosmic mass.

Exclusively concerning what it means for science, the birth of the universe revealed by quantum physics and the astrophysicists conflicts with the conception about black holes and the theory of relativity proposed by Einstein, insofar as the general theory of relativity covers the very big things, while the universe is smaller than an atom. More precisely, it is at small scales that the key incompatibility appears, since the notion of a smooth spatial geometry—the basic principle of general relativity—is destroyed by the violent fluctuations of the quantum world at minimum distance scales. This essentially means that general relativity and quantum mechanics "cannot both be true at the same time," that science is torn between one vision and the other and that, in short, it needs to find a harmony between the two of them, a quantum theory of gravity. The superstring theory has apparently solved the tension existing between these two theories, showing that they need each other and that the marriage between the laws of the big and the small turns out to be inevitable as well as necessary (Greene, 2006, pp. 15–17 and 192).

The Childish and Youthful Nature of Modern Art

In relation to art, Charles Baudelaire draws attention in *The Painter of Modern Life* (2004, pp. 82–91) to the barbarian or childish character of modern artists. He writes with regard to this that the root of their genius lies in their curiosity and in their "convalescent" state, which suggests a return to childhood.[3] After all, convalescent patients—like children—enjoy a vital interest in everything, joyfully absorb shape and color, and, due to their profound curiosity, watch everything as something new like someone who is always "inebriated." Therefore, the modern artist is a man-child, and the genius represents a recovered childhood, albeit endowed with a spirit which allows him to involuntarily organize, classify, and harmonize the materials accumulated in his memory. In effect, all these materials undergo an obligatory idealization, which results from a childish, sharp, deeply naïve, and accordingly magical perception.

Nevertheless, if modern art relates to childhood, it does to youth too and, in this context, a special mention needs to be made of the significant theories, practices, and changes operated, particularly by Artistic Avant-gardes, between the beginning of World War I in 1914 and the end of World War II in 1945. They initiated a new stage, driven by the two world crises and by an unprecedented collective anguish in history. Hence, this is perhaps why their essential characteristic is a questioning of the self and of their time, as well as the transformation of the art system or structure from a representative to a functional one (Argan, 1998, p. 281). In my view, this denotes that the aim was no longer to represent a world that raised a strong unease, but rather to stress the action, what to do and how to do it, all possibly for the purpose of transforming what has been given to us. Next to this stands the idea of renovation, rejuvenation, and novelty, defined by the opposition to the old and to the preterit. However, as will see later, the(ir) drama lies in the fact that their flight from the past and, somehow, also from the present consciously leads to an evolution without a future, without a project, and without sense (Subirats, 1989, pp. 84 and 107–108).

---

[3] About the childish character of the creator, it can be seen (Neumann, 2004, p. 36; Gardner, 2010, p. 51 ff).

In any case, prior to making a critical tour over time through the most outstanding currents and artists of avant-gardes, along with their most important postulates, it is worth highlighting that, broadly speaking, three "young" rebellion paths can be established. One of them, cubism, implies the culmination of a new process when looking at art and at the world which originated in impressionism, going through Cézanne, and which sets the path toward abstraction. Another one leads to surrealism, to lettrism, to situationism, and, later on, to student revolts, and to some terrorist explosions, whereas the last one, after M. Duchamp, gives rise to contemporary music, neo-dadaism, pop art, and the different artistic practices that appeared in the second half of the twentieth century in the United States. As a matter of fact, these three trails may be summarized in two major attitudes of contemporary art: "violence and humor, the anarchic shout and the quietist whisper, dissatisfaction and acceptance, provocation and indifference, the transgressor gesture and the everyday gesture" (Watson, 2002, p. 77; Granés, 2011, p. 46).

The antecedent of Artistic Avant-gardes can be found in the cubism of 1907–1908, a key artistic current, since it arises as the basic axis of twentieth-century art. If Cézanne's great works—in turn, the immediate precedent of cubism—were created in the same months during which Einstein prepared his theories, the whole change that art was experiencing constituted a reflection of the scientist. Not in vain, the history of avant-gardes—similar to that of relativistic physics and quantum mechanics—can be conceived as a purification process in search of what is essential to it (Danto, 1999, p. 87); and that explains why they share a common interest of the search for fundamental units, for that deeper reality from which the new forms could emanate (Watson, 2002, pp. 77–78). This is what happens, for instance, in cubism, which tries to deepen into things and highlight structure instead of appearance, and in early-twentieth-century physics, for which the atom is more real than a fragment of visible matter (Barzun, 2001, p. 957).

The "futurist revolution," the first avant-garde movement strictly speaking—developed in Italy (1909–1931)—shows a cruel and violent art willing to sow a warlike spirit in the souls of Italians, additionally claiming that the entire Italian artistic heritage must be burned in order to be able to build the future. Meanwhile, New York (1908–1917) witnesses the first seeds of the "Dadaist revolution," characterized by rebelliousness and transgression, exactly as it happens in the "Dadaist revolution"—Zurich (1916–1920)—which imitates the savage and the child, rejects art, and vindicates life. Marcel Duchamp (1915–1923) in turn announces the "rebellion for oneself" in New York, and the musician John Cage (1937–1952) dilutes every distinction between art and life by claiming that the artist's task does not consist in imagining, fantasizing, or reorganizing reality for the purpose of generating a fictitious world or a harmony that seems artificial, but rather in exhibiting the world as it is and in teaching others to accept it. Following the war, in Zurich and in Berlin (1915–1918), art preferred to look at childhood and primitivism—as cubism had done—and it sought salvation and an escape from time, something similar to what the "surrealist revolution" was going to do a few years later (1924–1941). However, while Dadaists consider that art must cure, according to surrealists, it has to be a social revulsive. More precisely, surrealism believes that the West, a civilization that it despises alongside its Greco–Latin foundations, finds itself in its "twilight" and that its future is consequently doomed. In its stead, surrealism suggests collecting African masks—exactly as cubism previously did—as well as idols and figurines from Oceania and from North American native Indians. It thus makes an attempt to move away from Western civilization and from its bet on reason and—as with Romanticism—embraces irrationalism. An example thereof is P. Gauguin, for whom the Western white man no longer looks for pristine purity at the end of the world, but in primitives and barbarians. All of this can be explained by the fact that all these first postwar artistic movements are trying to rename the world destroyed by the conflagration and to save mankind, for which they wish to travel back in time, up to childhood or the primitive state, with which they seek to "refound" the world. This escape from time, which enables artists to create their own paradise, follows a variety of paths: an idealization of future society—Marinetti; a return to childhood—Ball; the inclination toward the primitive and irrational—Huelsenbeck; and a rejection of everything other than fun or life—Tzara (Granés, 2011, pp. 23–86).

In the 1950s, dissatisfied youth, chance, chaos, improvisation, spontaneity, and group collaboration came into play in the United States.

Around the same time (1948–1954) in Paris, youngsters' rebelliousness also arose as the new starring character in history. This is what happens in the "situationist rebellion," which, pushed by the fight against boredom, focuses on action in order to change life in its smallest details; therefore, it was all about creating not art or literature, but the "art of life." These same postulates were shared by other movements of the 1960s, such as lettrism and the hippy or yippy movements, mainly advocated by young people, the basic promoters of the May 1968 Revolution, which protests against welfare and show, full—in their view—of alienation and tedium. The May 1968 Revolution additionally achieves a number of successive transformations of everyday lifestyles, among which the following stand out: social changes in fashions, sex, hedonism, consumerism, the implosion of human rights, women's empowerment, and the recognition of ethnic, racial, and sexual minorities.

As for North Americans, there were two types of revolutionaries: those ascribed to the so-called psychological branch (that of peace) and those who joined the sociological one (that of fighting). The former seeks personal growth, spirituality, harmony, and creativity, and their representatives travel to India, Japan, or Nepal in the hope of finding such values there. In turn, those in favor of the second current crave for recognition, identity, and vengeance, which they hope to find in their travels, first to Cuba and later to China (Granés, 2011, pp. 111–217).

The latter did not manage to knock society down, though, their success being only partial, since neither bourgeoisie nor capitalism, despite these revolts, was ever at risk. Nonetheless, their proposals did soften hierarchies and taboos; they legitimized hedonism, came close to the margins of individual freedom, and opened up spaces for the expression of the self. In this sense, S. Sontag wrote in 1969 that "to attack the system is to do something for oneself, for my true self," which means that the transformation of society by art no longer consists in becoming involved in politics or in public issues, or even in the creation of utopias, but rather in driving the most radical individualism. Consequently—as M. Duchamp had already announced—art becomes associated with the notions of individual ingenuity, feeling, fantasy, and the invention of unprecedented rules, and transforms itself into an individual affair (Eco, 1972, p. 128; Granés, 2011, pp. 197–444).

3 The Loss of Prominence of Youth and the Aging of Society and Creativity

The Role of Wars in Social and Cultural Aging

The prominence of the transformative energy characteristic of childhood and youth which is so typical of the first modernity (particularly, as seen earlier, in science and arts, but also in the conception of democracy and the State, as will be shown later) will gradually give way to physical, cultural, and social decadence in the second modernity. And, without a doubt, a decisive importance corresponds here to the effects caused by the wars that hit the twentieth century, including the demographic aging of the Western population.

In its deepest roots, the senescence of our civilization, despite having historical origins that will not be dealt with here, probably has to do with the fact that our current form of civilization is heir to the destruction caused by the two most important armed conflicts that occurred in the twentieth century. During World War I, for instance, as the historian J. Barzun argues in *From Dawn to Decadence* (Barzun, 2001, pp. 1020–1043), the individuals who die are mainly the young men. As a matter of fact, the military service was compulsory in the nations involved therein (except for England), which means that every man above the age of 18 had to be mobilized; that age was even reduced to 16 at the end of the conflict. Thus, during the 52 months that the war lasted, some 10 million people died, while another 20 million were wounded. However, the irreparable loss that these figures imply has much more importance and a far greater scope than the actual death of soldiers, since the crippled ones, tuberculosis sufferers, those traumatized by war, the saddened ones, those driven mad, the ones who committed suicide, those with a broken spirit, with destroyed careers, the potential geniuses buried, or the births that never occurred all represent unfathomable losses at a human as well as social and cultural level. In this last sense, the failure on the part of civil and military authorities to coordinate their plans meant that plenty of artists, especially the young ones, succumbed in the trenches, or, expressed differently, that the group of most lucid artists, scientists, intellectuals, and eminent personalities from various fields was dramatically reduced by the war (Wohl, 1979; Stromberg, 1982; Isnenghi, 1989; Prochasson, 1993, ch. X; Prochasson y Rasmussen, 1996; Hanna, 1996; Hynes, 1992; Field, 1991).[4]

The same level of risk affected cultural infrastructures, that is, works of art, libraries, and the architectural constructions of interest—cathedrals, churches, monasteries, palaces, castles (Ziemann & Hofrichter, 2016, pp. 141–159). To this must be added the constant and intense anxiety suffered by millions of people in relation to the fate of their country, of their relatives, or of their friends on the frontline, which meant a permanent hindrance for a calm reflection. Furthermore, while fighting, soldiers felt that the end was near, which made them feel attracted by danger and death (Barzun, 2001, pp. 1020–1043).

In sum, it was a senseless waste of lives that deprived the Western postwar world of a large part of the most outstanding intelligences, of a whole young generation that could not only have questioned the old-fashioned views and consequently renewed society but also reinvigorated the necessary links with the culture that prevailed before the war (Barzun, 2001, p. 7). And, most importantly for this chapter—as highlighted by the historian J. Barzun (2001, p. 1048)—"[t]he *devastation*, *both* material and *moral*, had gone so deep that it turned creative energies from their course, first into frivolity, [which, as I mentioned earlier, has happened to the arts] and then into the channel of self-destruction."

A Demographically Aged Society

It seems obvious for social and cultural aging to be related to demography; after all, society has entered the final stage of Demographic Transition, which implies a new demographic regime full of transformations resulting from an improved quality of life, from technological innovations, and from the breakthroughs in medical science (neonatology

[4] Some of the most famous names of those killed in the conflict can be seen in the article in the newspaper *La Vanguardia*, from Barcelona, dated 04/12/2017. Seen, https://www.lavanguardia.com/hemeroteca/20170409/421524222390/primera-guerra-mundial-personajes-celebres-bajas-deportistas-escritores-poetas.html. Downloaded 23/05/2021.

techniques, antibiotics, drugs, prostheses, surgeries, etc.). This is how the conditions have been established for the creation of the "elder" as a construction of society and of its widespread aging. An unrepeatable and revolutionary phenomenon thus becomes manifest which has entailed a great efficiency in reproduction, insofar as most of the born individuals live until their old age, which is associated with overcoming that traditional radical qualitative threshold according to which the mortality that occurred between birth and the age of 15 was always situated around 50% and constituted the real reason for the extremely high fertility rates of the past. This has simultaneously triggered an accelerated increase of the population, since mankind has grown as a species in the last century five times as much as in the past, while fertility has at the same time gotten rid of a peremptory reproduction. To this must be added that the higher life expectancy and the demographic boom have taken place coinciding in time with a steady decrease of fertility to minimal values never seen before (Pérez Díaz, 2016, p. 2).

This change in the pyramid, known as "demographic aging," has turned into a multifactor and multifaceted issue which needs to be analyzed on an interdisciplinary basis, but in any case, it entails a growing weight of longevous people. For example, together with the progressive reduction of birthrate (it stands at 1.2 in Spain at present) and the lower mortality rate, substantial modifications have been operated regarding life expectancy at birth. Thus, the Neanderthal hominid had a life expectancy of 18 years on average which sufficed to perpetuate the species, though not to build an elaborate civilization, and even in the early twentieth century, life expectancy only reached 34 years in Spain. It has spectacularly increased today, in such a way that prolonged old age, shaped by the "old young ones" (60-to-75-year-olds) and the "old old ones" (from the age of 75 years onward), is now experienced by millions of people, among them octogenarians. Therefore, 64+ people currently account for 15% in developed countries and, in Europe, 33% of the population will be old around 2050, whereas the groups ranging between the ages of 15 and 64 will become smaller. As for Spain, 17% of its population is 65 years or older at present, a segment that will come close to 30% by 2050 (Hidalgo González, 2001, pp. 4–65; Fernández et al., 2010; Pérez & Abellán, 2010).

This aging of the Spanish population is clearly exemplified by the empty or semi-abandoned villages of inland Spain, the expensive economic cost of pensions (a real burden for the competitiveness of enterprises and of the country as a whole, for the sustainability of the welfare State, and for the actual youngsters who work and pay contributions), the collective overexertion to preserve life, and doing so with quality and a minimum dignity, of so many people who find themselves on the brink of death or when their degree of disability increases, to maintain their own existence (Pérez Díaz, 1998, pp. 2–10).

It is true that the institutions, the old individuals affected, and society itself try to counter this overall process of aging—and of creative decay—in an attempt to make it possible for them to live longer and better. Neither should we forget that the life cycle has been transformed, and the significance of each life has become different as a result. On the other hand, the new seniors are urban, have gone to school, own personal assets, and are financially solvent, in addition to being in good health condition. Hence, society fosters an active aging which tries to boost people's abilities when they grow older; in other words, they encourage them to have a productive life in society and in the economy. The term "active" implies that these individuals should participate in social, economic, cultural, spiritual, and civic affairs, and not only in physical or work-related ones. In a nutshell, the aim is to ensure that the different stages of life, and those associated with the old ones, do not become closed and stagnant cycles (Pérez & Abellán, 2010).

In any case, this program for the rejuvenation of seniors is still in the making, or, expressed differently, it constitutes an ideal rather than a reality put into practice, which means that we still do not know whether it is possible to prevent the aging of society, at a spiritual, cultural, economic, and even political level.[5] Therefore, the senescence process undergone by contemporary society is evident, and so is the loss of young people's power of conviction—and of transformation. Demographic aging does not appear as the main reason for this, though; instead, it is just another expression of a dilated social transformation process due to which young

[5] Needless to say, coronavirus is changing the perception of old age, but that is not going to be analyzed now, since we must wait and see what happens in the near future.

individuals have stopped playing the central role that corresponded to them a few decades ago (Pérez Díaz, 1998, pp. 2–10).

The Aging of Present-Day Art and Science

Despite not being the only causes, the disastrous effects of wars and the gradual aging of the population are behind the process through which the early-twentieth-century energetic creative moment faded little by little until it fell into a clear decline of arts, sciences, and of society itself in the second modernity. As for contemporary art, it is worth wondering what general assessment can be made about the group of the two artistic avant-gardes, as well as asking ourselves what their failures were, and what contribution they eventually made to society. First and foremost, it deserves to be mentioned that the assessment is not completely negative, insofar as artistic avant-gardes cannot be said to have proved useless in the long run. After all, in contrast with the evident collapse and failure of communism's social and political revolution, the cultural contribution of artistic avant-gardes has been really significant and still remains current (Granés, 2011, p. 86). This undoubtedly relates to the fact that they were assimilated by bourgeoisie, the most powerful social class at the time. Instead, the scientific innovations, which served as a model for artists, were not understood by that social class, since the most outstanding European scientists of the period—Einstein, Rutherford, Ehrlich, and Baekeland—had a solitary, self-sufficient, and self-absorbed character in common. In fact, science did not receive any information from art, whereas the latter—as I mentioned earlier—did benefit from scientific knowledge (Watson, 2002, pp. 122–123).

In any case, beside these positive aspects, the first avant-gardes have other more negative features too. It is worth highlighting in this regard that second modernity artists no longer believe in originality, they defend what lacks a beginning, and, therefore, only manage to safeguard the appearance of innovation (Sloterdijk, 2019, pp. 227–229). Furthermore, the supposed search for originality was limited to an initial moment because, shortly after, it became a myth, insofar as that originality actually emerges as an active result of repetition and recurrence. Finding the reticle, for instance, provides artists with a sense of beginning, of a new start, of a zero degree, of a rediscovery, and in a nutshell, of finding themselves in front of their origin. However, the reticle does not endow artists with originality, but forces them to repetition and, consequently, this experience of the original, which has affected generations of artists, critics, and audiences, not only is false but ultimately constitutes a fiction; the origin of artistic modernity accordingly dissolves in an endless replication (Krauss, 1996, pp. 171–182). Nonetheless, the reason for this to have happened lies in the fact that the prevalence of reproducibility has evolved to become the essential postulate of avant-garde aesthetics and reproduction has turned into the principle of industrial and commercial artistic creation, which put an end to the original and utopian moments that defined the pioneers of modern art (Subirats, 1989, p. 21).

On the other hand, the passage from cultural to political terrorism marked the end of "the first time of cultural revolution" in Europe and, since then, both the most transgressor art and terrorism have needed to transform their actions and practices into a show as a prerequisite to be successful. On these grounds, the rebels of the "second period of cultural revolution" stopped fighting against the system and decided to enthrone the show, which also affects the actual myth of counterculture, since the latter, by becoming one of the most profitable entertainments of cultural capitalism, came to its end as well (Granés, 2011, pp. 298–326). Certainly—as pointed out by Arthur C. Danto—present-day art has a deeply plural and tolerant nature, since it follows no rules and is consequently unpredictable; and it ultimately represents the very embodiment of human creativity and freedom (Danto, 1999, pp. 20–139). Nevertheless, we should reflect on what this nonnormative, uncertain, and unexpected art has really become and, above all, what distinguishes it from what it is not.

One can actually state that the monumentalization of postmodern architectures which seeks certain metaphysics of the sublime, along with the cult of power that they express, has made it clear that this last stage of modernity arises as an epigonic and decadent movement. In effect, the "second cultural revolution" becomes frivolous and banal, as exemplified by A. Warhol; avant-garde is "exhausted," in the words of G. Lipovetsky, and ends up having neither a project nor a goal (Granés, 2011, pp. 298–326), without a narrative, and without originality as an original origin. This explains why avant-garde is ultimately led toward its end, or rather toward the end—already announced by Hegel—of a long historical period and an important social function.

In the end, it comes as no surprise that "in its cultural, stylistic, and artistic emptiness, the rupture, the protest and the overcoming advocated by avant-gardes—paradoxically—signals the obsolescence of avant-garde itself, that is to say, its death" (Subirats, 1989, pp. 21 and 159–160).

Probably, at the basis of it all, at the root of the accelerated search for pleasure favored by contemporary creation—and particularly, fashion—stands the intensification of the sense of earthly transience, of the sorrow for aging, of the nostalgia for youth or, expressed differently, the melancholy of time and the anxiety associated with leaving life. Not in vain, as Charles Baudelaire wrote in *The Painter of Modern Life* (2004, p. 123), "fashion must (…) be regarded as a symptom of the taste for the ideal which floats in the human brain above all the vulgar, mundane, and unclean things that natural life stacks in there, as a sublime deformation of nature, or rather as a permanent and successive attempt to reform nature."

It is true that, at the foundations of this reform lies nothing but the essential anthropological paradox shaped, on the one hand, by the awareness of aging and, on the other hand, by the desire to stop that process and to rejuvenate, that is, the desire not to be nature, to survive, not to die. Note, though, that the strong resistance of the economic system has halted the momentum toward rejuvenation and has somehow also slowed down the desire to survive.

Finally, although science has kept generating new inventions and approaches throughout the twentieth and twenty-first centuries, thus making new knowledge and utilities known to us, I am not sure whether they fit the perspective of creativity that I am trying to unveil here, or, to put it in another way, whether they can help build a narrative original creativity. The science historian Sánchez Ron (2000, pp. 294–297) refers in this context to the "last scientific revolution of the century," which comprises genetics, molecular biology, biotechnology, and genetic engineering; despite being considered sciences, they also largely appear as pure technology, since their development inevitably relies on a great wealth of information—high-speed computers and systematic methods to distribute, organize, and classify all that information. It is worthy of note in relation to the previous discussion that biological sciences often describe substances as transmitting "information"—the neuron speaks to the chip and this in turn speaks to the nerve cell—and refer to the existence of "codes" that regulate such exchanges (Barzun, 2001, p. 1108).

Likewise, the time of great conceptions which revised the traditionally accepted visions is now over, without forgetting that, inside science, technology stands out for being the only institution which has undergone no reductions of results. To this must be added that the common sense which prevailed until now no longer governs science (Bronowski, 1978, pp. 105–129) and that the scientific reports on highly diverse issues are often contradictory, as it happens, to quote but a few, with those related to global warming, the presence of radon on the soil, the agent orange, food additives, genetic manipulation, or the Covid-19 virus, all of them topics on which it becomes really difficult to develop an intelligent homogeneous opinion. This explains why society no longer trusts science as it used to in the nineteenth century, why society and technology have lost their sanctity, and why they are not free from strong social and philosophical criticism which holds them accountable for the worst ills of the present. Among these are that they have helped destroy the spiritual part of human beings due to an excess of rationality and mechanics; or that they have aroused fear of the physical annihilation that nuclear weapons can cause, or the mental disorder resulting from genetic manipulation, or the unease raised by cloning due to its ethically dubious procedures (Barzun, 2001, pp. 1109–1177).

It is in all these senses that this type of creativity embodied by today's science does not fit within the coordinates addressed here. Moreover, neither can it be compared in terms of relevance with the one shaped by relativistic and quantum physics, which, as seen earlier, were truly revolutionary, rebellious, and youthful, apart from favoring the development of the subsequent science.

4 From the Childhood and Youth of the First Modernity to the Aging of the Second One

The Infantilism and Youth of Creators and the Generational Conflict

As shown in this chapter, the origin of creativity, which is inherent to life itself, to creative living, to the condition of being alive, which accompanies human beings from birth to death, settles during childhood; what is more, infantilism stands out as one of the defining features of creators. More precisely—according to Psychoanalysis—creation works arise from mothers' realm and set in motion a process to recover from the child trauma experienced after leaving the mother, which is why the past suffering of creators lies at the base of their creation.

Nevertheless, from my perspective, creativity cannot be exclusively understood—as Freud and an important sector within Psychoanalysis suggest—as a subtraction, as a response when faced with the absence of something, or before suffering or unhappiness, but also as a very human wish to be happy, to add, to transcend our nature, to increase one's own life, to make more out of life, in the Simmelian vision; a desire to enable life to have a more significant scope and higher intensity. This becomes possible through an energy which repairs existence, which could explain why creativity always represents a return to the origin and why, from the latter, creative originality flows with renewed effervescence, and, therewith, the establishment of a new universe, of a world that always seeks to improve the imperfection that is inherent to it.

Together with childhood, youth is the most creative biological stage, the most tempestuous and innovative one which has novel and fresh feelings, along with the ideas for a new life; and that is why youngsters represent the picture of changing times. However, they achieve it through the clash between young people's cosmovision and that of society itself—and, in particular, the one of seniors. Hence the key role played by the conflict between generations—especially addressed by Mannheim and Ortega. My focus here is placed on what the concept "generations" means, on what role is assumed by youngsters in them, and on how they work and become linked to creativity. Most importantly in this respect, a generation relies on the biological rhythm of birth and death, which entails a generation clash of creators both with themselves and with their society, additionally emerging as a fundamental notion within social historical evolution. After all, generations overlap or combine and therefore come into conflict, thus becoming the core of social coexistence and of its dramatic dynamism. In fact, two types of generations exist: those of youngsters who fight against the old ones—this is what basically happens at the beginning of the first modernity; and those of young people who coexist with their elders, toward whom the former show their solidarity and to whom they are subordinated—as illustrated by the aged second modernity where youth has lost its previous prominence.

The dialectics triggered by this conflict between generations is usually displayed as follows. The generation that forged the normative culture—exemplified in the artists of the first avant-gardes and in the most important relativistic and quantum physicists—grows old, "dies," and retires at the age of 60, which leads to the oblivion of that culture by the new generations, something that in turn becomes necessary for the usual social evolution derived from the fight between the young and the old. Nonetheless, creativity constantly reminds us that social memory is as essential for society as its historical loss (Mannheim, 1993, p. 213); hence the dialectics between the origin and the original staged in it—the main reason for writing this essay—seems appropriate to express the generational fight between the adults who receive, live, file, ritualize, and protect their ancestors' collective memory, the foundation of social ties (Halbwachs, 2011, pp. 25 ff. and 99 ff.), "the suitable alternative of history"[6] (Ricoeur, 2010, p. 158), and the young ones who want to erase it to create a new one of their own.

[6] Paul Ricoeur (2010, p. 157) defines collective memory "as a selection of marks left by the events that influenced the course of history for the groups concerned; and it is recognized to have the power to stage those shared memories on the occasion of festivities, rites, and public celebrations."

The First Modernity Constitutes a Stage of Childhood or Youth

This conflict between generations became apparent in the first modernity, particularly between 1905 and 1908, a highly creative stage which generated four innovative forms of energy: praxis and the devotion for violence; major artistic successes; scientific innovations; and social reforms, more specifically the conceptions of State and democracy. Broadly speaking, these four inventions have arguably gone from a rejuvenation stage to one of decadence or aging.

Indeed, the first modernity as a whole can be regarded as a childish or teenage phase and, accordingly, as an immature one where its "killer" nature—the terrible modern wars and sociocultural upheavals—stands out alongside its creative dimension, especially in science and art. That is why young people's ideas, theories, and postulates succeed in defeating those of the old ones, thus shaping a rejuvenation and revitalization stage within society.

According to Einstein's approach, space and time become not only relative but also "moving sands," which eventually serves as the basis of the subsequent "world without time" in which the present no longer exists (Rovelli, 2018, pp. 33–72 and ff.) and where everything vanishes (there is only one step from there to the "con-fusion" of the present). Furthermore, the discovery of the *Big Bang*—resulting from the joint advances of relativistic and quantum physics—implies a yearning for the return to the origin that is characteristic of the modern period and, in parallel, a legitimization thereof via science. Thus, most significantly, the expansion of the universe that begins and seemingly finishes with a final explosion (a *Big Crunch*), with an apocalyptic sort of deflagration, denotes the mythical footprint left by the biblical books *Genesis* and *Apocalypse* on this scientific narrative of the universe. However, if we think about it carefully, this expresses another—indirect—manner of nostalgia for paradise, of yearning for originality. Therefore, it is worth highlighting that the latter entails a new reflexive recreation of the origin, with an elaborate mathematical and geometric—and consequently abstract—content, aimed at thinking, imagining, and building the future.

Evidence has also been provided about the barbarian or childish character of the modern artist, who is a man-child, whereas the genius represents the recovered childhood. Likewise, modern artistic praxis tends to be associated with the rebellious or revolutionary youth that—essentially—tries to rename a world devastated by the world wars, for which purpose it travels back in time, as far as childhood or the primitive state of mankind, in an attempt to "refound" society.

What these creations—physical science and the arts (outstandingly cubism)—have in common is the pursuit of a deeper reality, from which it can be inferred that it also constitutes the foundation, the origin of the creative process.

The Widespread Physical, Cultural, and Social Decline of the Second Modernity

The regenerative energy of childhood and youth which characterizes the first modernity gave way to a widespread physical, cultural, and social decline in the second modernity. A key role in this transformation corresponds to the demographic aging of the Western population and to the wars that ravaged the world during the twentieth century. Both phenomena therefore go hand in hand with the senescence of the arts, science, democracy, the 15-M movement in Spain, and capitalism.

As for demographic aging, it favors this widespread social decay, insofar as it brings about a whole series of first-rate consequences at a human, social, cultural, and economic level, and, of course, within the field of creativity. In this last sense, it specifically causes a considerable loss of collective vitality derived from the absence of the youth that promotes creativity and entrepreneurship, as well as the correlative threat of paralysis, decadence, and/or extinction. It additionally buffers young people's desire for rebellion, since there is a generational coexistence with old people who make closer links possible through filiation lines and the development of more "vertical" ties associated with relationships, support, advice, exchanges, and affection (Pérez Díaz, 1998, pp. 2–10). Youngsters are consequently not so keen to get rid of the postulates of their elders in society or to oppose them.

It is as if the energy displayed by childish creators full of youthful rebelliousness had been exhausted, as if child and young creators, and their ideal, had surrendered—or had been forced to do so—to the plans of the system established by their elders.

Both world wars, which entailed the massacre of millions of youngsters, among them the best-trained and most creative ones, have also played a role in this decline of creativity. Thus, due to the commotion caused by those armed conflicts, creativity leaves its course (its time narrative, I would say, in keeping with what has been explained in this chapter), as though history had succumbed, as if culture could not help feeling attracted—as with young soldiers—by a passion for the end which replaces the one for the origin (Roche Cárcel, 2007, p. 37 ff.) or as though it became frivolous and self-destructive.

In effect, because of the impact caused by war, creativity has stopped being understood as deconstructive or destructive, and that can easily be seen in literature and in the arts. For instance, Apollinaire, after coming back from the front wounded, changes his techniques and returns to the nineteenth century; Stravinsky finds his inspiration in Pergolesi; Léger and Picasso focus on representation, after cubism; Duchamp becomes the standard-bearer of counter-modernity, thus finishing a long stage of art construction and initiating that of destruction (Barzun, 2001, p. 1062). In this context, it only seems logical for the first modernity to have been followed by another one ready to be matured by the requirements of society itself and by its desires for hedonism and pacification, being fed up with so much barbarism. Late modernity will therefore be less prone to big upheavals—even though it will still witness regional conflicts, many of them inherited from the first modernity—and will be dominated to a greater extent by adults, since it is a society, and a civilization, on its way to decay or aging.

The phenomenon was certainly anticipated in the early twentieth century, when young intellectuals worshipped a whole series of literary and artistic works that drew attention to the decadence of society (*The Wasteland*, by T. S. Eliot, *Ulysses* by Joyce, *Heartbreak House* by Shaw, and *In Search of Lost Time* by Proust). At the same time, society enters a large-scale primitivism phase in which "ridicule, denial, antiart, and sensorial simplicity" precisely denote that culture and society are falling into decline. Furthermore, the fact that intellectuals sought novelty and originality ultimately brought about the overlapping of several styles, which likewise suggests that the period could no longer have a characteristic or canonical style—the average lifetime of a style at the end of the twentieth century was three months. In other words, the impetus born in Renaissance had become exhausted, and the new start which took place in the years prior to 1914 had come to an abrupt halt (Barzun, 2001, pp. 1053–1058).

Moreover, "before the moral, sexual, and political rebellions of the 1960s, the attack against authority, the derision of everything that was established, the distortions of language and objects, the indifference towards clear meanings, the violence against the human form, the return to sensorial elements, and the increase of 'anti' genres turn modernity into a mirror of disintegration and of the incitement to delve deeper into it." It deserves to be highlighted in this regard that art joins science and technology—especially in cubism—in the love for the speed of cars and planes, as well as in the idea that the most recent is the best. This explains why the modernist looks forward and is an innate futurist, why he challenges the old conviction of traditional wisdom and the prudent value of conservation. However, it follows from this not only that all the old things are obsolete but also, ultimately, that artistic avant-garde itself will very soon become obsolete, old. And added to all this, the subsequent art moves along the waste—the sea waste collected on the beach—the garbage or the disposable items, fragile and ephemeral materials, and the dehumanization of the human form, which means that art as an institution, with a moral or social purpose, is dead (Barzun, 2001, pp. 957 and 1051–1070).

The artistic assessment is simply devastating and could be summarized by saying that, in Modernity, we have gone from "the Great Hope" to "Civilization must disappear" (Barzun, 2001, p. 1003 ff.). After all, the origin pursued by avant-gardes ends up dissolving in a mechanized, repetitive art, in an endless replication, with no meaning, with no objectives or goals, and, accordingly, with neither a past nor a future. However, the maximum creativity of late modern art, freed from styles and norms, does not seem to revolve on elucidating what its foundation—or its project—is, or, in short, on what it has been transformed into, where it is heading, and what contribution it makes to contemporary society.

In turn, science has largely been transmuted into technology, it has stopped producing big theories meant to replace the old physics, has lost a considerable part of its legitimacy, filling society with fears, and has become more opaque, more elitist, and less transparent, thus worsening social differences.

Representative democracy is not at its best either. Its internal and external enemies have increased both quantitatively and qualitatively, and it has been deprived from most of its legitimacy—like science—while simultaneously witnessing an increased ideological and social polarization, instead of finding the equilibrium that it achieved through a large part of modernity.

Bibliography

Argan, G. C. (1998). *El arte moderno. Del Iluminismo a los movimientos contemporáneos*. Akal.

Barzun, J. (2001). *Del amanecer a la decadencia. Quinientos años de vida cultural en Occidente (de 1500 a nuestros días)*. Taurus.

Baudelaire, C. (2004). *El pintor de la vida moderna*. Colegio Oficial de Aparejadores y arquitectos Técnicos, Consejería de Educación y Cultura y Fundación CAJAMURCIA.

Bauman, Z., & Leoncini, T. (2018). *Generación líquida. Transformaciones en la era 3.0*. Paidós.

Bell, D. (1992). *Las contradicciones culturales del capitalismo*. Alianza.

Bronowski, J. (1978). *El sentido común de la ciencia*. Península.

Chown, M. (2007). *El zoo cuántico. Guía turística del interminable Universo*. La liebre de marzo.

Danto, A. C. (1999). *Después del fin del arte. El arte contemporáneo y el linde de la historia*. Paidós.

Eco, U. (1972). *La definición del arte*. Ediciones Martínez Roca.

Elias, N. (2013). *Sobre el tiempo*. FCE.

Fernández, J. L., Parapar, C., & Ruiz M. (2010). El envejecimiento de la población. *Lynchos. Cuadernos de la Fundación General del C.S.I.C.*, No. 2. Edición digital.

Field, F. (1991). *British and French writers in the First World War*. Cambridge University Press.
Gardner, H. (2010). *Mentes creativas: una anatomía de la creatividad*. Paidós.
Granés, C. (2011). *El puño invisible. Arte, revolución y un siglo de cambios culturales*. Taurus.
Greene, B. (2006). *El universo elegante. Supercuerdas, dimensiones ocultas y la búsqueda de una teoría final*. Drakontos bolsillo.
Halbwachs, M. (2011). *La memoria colectiva*. Miño y Dávila.
Hanna, M. (1996). *The mobilization of intellect. French scholars and writers during the Great War*. Harvard University Press.
Hidalgo González, J. G. (2001). *El envejecimiento. Aspectos sociales*. Editorial de la Universidad de Costa Rica.
Hobsbawm, E. (2001). *Historia del Siglo XX*. Crítica.
Hobsbawm, E. (2013). *Un tiempo de rupturas. Sociedad y cultura en el siglo XX*. Crítica.
Hynes, S. (1992). *A war imagined. The First World War and English culture*. Pimlico.
Isnenghi, M. (1989). *Il Mito della Grande Guerra*. Il Mulino.
Krauss, R. (1996). *La originalidad de la Vanguardia y otros mitos modernos*. Alianza.
Mannheim, K. (1993). El problema de las generaciones. *REIS, 62*, 193–242.
Martin, M. (2008). La teoría de las generaciones de Ortega y Gasset: una lectura del siglo XXI. *Tiempo y espacio.*, Year 17, *20*, 98–110.
Neumann, E. (2004). El hombre creador y la Transformación. In E. Neumann, M. Eliade, G. Durand, H. Kawai, & V. Zuckerhand (Eds.), *Los dioses ocultos: Círculo de Eranos II*. Anthropos.
Ortega y Gasset, J. (1983). *El tema de nuestro tiempo*. Alianza Editorial/Revista de Occidente.
Ortega y Gasset, J. (2008). *En torno a Galileo*. Revista de Occidente-Alianza.
Pérez, J., & Abellán, A. (2010). Sociedades longevas: un desafío para el Siglo XXI. *Lynchon. Cuadernos de la Fundación General CSIC*. Digital Edition No. 2.
Pérez Díaz, J. (1998). La demografía y el envejecimiento de las poblaciones. In A. S. Staab & L. C. Hodges (Eds.), *Enfermería Gerontológica* (pp. 451–463). McGraw Hill.
Pérez Díaz, J. (2016). El temor al envejecimiento demográfico. In J. S. Humet et al. (Eds.), *Edades en transición. Envejecer en el siglo XXI* (pp. 44–54). Ariel.
Prochasson, C. (1993). *Les Intellectueles, le socialisme et la guerre, 1900–1938*. Seuil.

Prochasson, C., & Rasmussen, A. (1996). *Au nom de la patrie. Les intellectuels et la Première guerre mondiale, 1910–1919*. La Découverte.
Ricoeur, P. (2010). *La memoria, la historia, el olvido*. Trotta.
Roche Cárcel, J. A. (2007). De la nostalgia por el origen a la pasión por el final. De la génesis de la sociedad al fin de lo social. In J. A. Roche Cárcel (Ed.), *Espacios y tiempos inciertos de la cultura* (pp. 37–68). Anthropos.
Rovelli, C. (2018). *El orden del tiempo*. Anagrama.
Sánchez Ron, J. M. (2000). *El siglo de la Ciencia*. Taurus.
Sloterdijk, P. (2019). *En el mundo interior del capital. Para una teoría filosófica de la globalización*. Siruela.
Stromberg, R. (1982). *Redemption by war: The intellectuals and 1914*. The Regents Press of Kansas.
Subirats, E. (1989). *El final de las vanguardias*. Anthropos.
Watson, P. (2002). *Historia intelectual del siglo XX*. Crítica.
Winnicott, D. W. (2013). *Realidad y juego*. GEDISA.
Wohl, R. (1979). *The generation of 1914*. Harvard University Press.
Ziemann, B., & Hofrichter, A. A. (2016). La violencia como objeto de estudio en las investigaciones recientes sobre la primera guerra mundial. *Historia Social, 84*, 141–159.

Part III

Creative Categories of the Originary

With Chap. 6, I begin the analysis of the creative categories of the originary, that is, of those that are theoretically directed toward the future. Here we will see the evolution of capitalism from the first to the second modernity, the first being characterized by creative youth, while the second is defined by its aging. Early modern capitalism possessed an intense creative and transformative energy. However, it was not the technical invention, but the social one, that was the main intellectual motive of the industrial revolution. In fact, in these initial moments, capitalism is cultural because its economic activities are guided, designed, and made meaningful to the participants through shared symbols, norms, beliefs, and values. In addition, the stimulation and satisfaction of new needs through industrial production was possible thanks to the erosion of the traditional attitudes of the workers, although, obviously, the role of the entrepreneur is also fundamental, not only limited to competing, but also transforming the nature of the competition. Therefore, with capitalism, it is not specialization that stimulates economic growth, but innovation, and its creative nature results. Likewise, the social relations to which the capitalist system gives rise, and not the means or the forces of production, constitute an expression of the creative capacity of humanity, that is, of human work and its general meaning.

In the second modernity, the capitalist economy has gone into decline, crisis, or senescence. The origins of this cataclysm lie precisely in the utopian effort of economic liberalism to establish a self-regulating market system, but the type of individualism that enthroned the capitalist system and that placed individual existence in a fluid context of permanent change also had its influence. It is now governed by the increasingly urgent and necessary rhythms of technical innovation. Thus, no tradition or habit can be consolidated and, with them, neither can any collective universe nor a stable framework of references and symbolic values, nor any historical memory that ensures the meaning and expression of the authentic human experience.

In conclusion, current capitalism is in a phase of senescence, to the extent that the old capital today becomes an important driver of technological innovation, but not of social innovation, something that had been the main intellectual motivation of the Industrial Revolution. Added to this is the fact that the capitalist economy has detached itself from society and has come to dominate it, while the traditional social cut of capitalism has been broken.

The enthronement of creativity as an emerging force in the field of Social Sciences, its empowerment as an almost absolute energy, could lead us to think that it would constitute a useful response to overcome the aging of capitalism, the labor crisis, and even to give prestige to this activity. The objective pursued by Chap. 7 is therefore to confirm whether creative work has achieved these objectives; if it really has persecuted them; and whether, in addition to generating large and rapid benefits for the creators and companies that have hired such work, it has become a powerful force both in social cohesion and in the globalization of the economy, society, and democracy and its values.

To develop the aforementioned objectives, this chapter has been divided into four sections. The first gives a general introduction, followed by Section II, where I will try to define work and frame it within the negative perception that the West has traditionally had about it. Additionally, the two models through which our civilization strives to provide employment are synthesized, and finally the characteristics of the work at present are summarized. Section III focuses on creative work in the context of the global economy and, within it, emphasizes defining what creativity and the global economy are, describing the most essential characteristics of cities and their creative industries, and how the latter are causing new inequalities. Finally, Section IV offers some final reflections designed to analyze the main features that creative work currently assumes.

As has been seen, the general approach of the book is dedicated to the character of originality that every creative process must have and how the original is linked to the concepts of origin and originary, with nostalgia for the past and the desire to construct the future. Chapter 8 ends the part dedicated to the categories of the originary. The aim is to analyze whether the global economic system has commercialized and de-tradized creativity, emptying it of meaning and collaborating, in this way, in the deepening of the crisis of representative democracies and their values. To do so, they will first have to be defined and the concepts of "crisis," "representative democracy," and "creativity" be interrelated.

It will be seen that the creative economy has created new inequalities between rich and poor and between creative and noncreative states, widening the gap between market and state, between politics and economy, between capitalism and representative democracy. And it has even activated the contradictions of capitalism, that is, it has deepened the crisis, the decadence, and the collapse of itself and of democracy.

In this sense, creativity cannot today be seen only as a lifeline for the many difficulties and uncertainties that plague representative democracies. It is also, and should not be forgotten as, a source of intensification of these same problems, while generating new ones that are not foreseen.

6

From Creative Capitalism to Old Age

1 The Creative and Transformative Energy of Capitalism in the First Modernity

Bourgeois business capitalism, which rationally organizes free work, arises only in the West. More precisely, Protestant puritanism assumes as its key task the rational domination of the world and its transformation, due to the fact that supramundanely oriented puritan rational ethics implements intramundane economic rationalism with all its energy and consequences. The paradox caused by this is that the work focused on this world looks somewhat distant to the puritan but simultaneously represents his interest in a transcendent purpose (Weber, 2017, pp. 338–405). In any case, the rationality that I am referring to—which is additionally typical of modernity—has an inextricable connection with natural science, inventiveness, and work organization (Jaspers, 2017, p. 158).

Well, capitalism arrives unannounced and by surprise. Indeed, nobody had anticipated that a machine-based industry was going to be established (Polanyi, 2017, p. 150). However, it was not technical invention but the social one which emerged as the main intellectual motive of the industrial revolution (Polanyi, 2017, p. 179). As a matter of fact, during

J. A. Roche Cárcel, *Creativity and Time: A Sociological Exploration*, Palgrave Studies in Creativity and Culture, https://doi.org/10.1007/978-3-030-84838-5_6

those initial moments, capitalism was cultural because its economic activities were guided, designed, and endowed with meaning for participants through shared symbols, rules, beliefs, and values.

Furthermore, stimulating and satisfying new needs by means of industrial production was possible thanks to the erosion of workers' traditional attitudes (Ingham, 2010, pp. 41 and 76). Obviously, a key role equally corresponded to the entrepreneur, who not merely competes but also transforms the nature of his competence. He thus constantly reforms the system by structuring the production factors so that they can bring benefits to the innovator, either by means of an increased productivity or through the manufacturing of new goods (Ingham, 2010, p. 51). In the case of a more explosive growth—as pointed out by Schumpeter—capitalism even fosters innovation, which consists in the ability to improve what already exists to make something new. Therefore, it is not specialization that boosts economic growth in capitalism, but rather innovation (Florida, 2009, p. 74); hence its creative nature.

Likewise, the social relationships triggered by the capitalist system, and not the means or forces of production, constitute an expression of mankind's creative capacity, that is to say, of human work and its general sense (Ingham, 2010, pp. 32–33). Not in vain, political economy is the one which relies on the ability to produce and, at the same time, to create values owned by us humans (Boada, 2017, pp. 61–62).

Summing up, the creative nature of the capitalist system can be essentially explained through the tension between two opposed movements that exist inside it. On the one hand stands out the idea supported by economic liberalism of establishing a self-regulated market that "implied an absolute utopia" based on the support of merchant classes which profusely used the *laissez-faire* and free trade methods. In effect, market economy arises as an economic system that is monitored, regulated, and governed by market prices, and this also entails the presence and circulation of money, which acts as purchasing power in its owners' hands. Alongside money are labor and land, whose prices will be respectively called "salary," "income," and "interest." Actually, this utopian character of market liberals—as Schumpeter sees it—leads them to invent the gold standard, a mechanism which, according to the economist's ideological conviction, would generate an increasingly prosperous, borderless world

(Polanyi, 2017, pp. 47–129) which additionally results in a permanent expansion of the market.

Nonetheless, the creativity of the economic system also stems from a second contrary movement which, in addition to being essential to protect society, is grounded on the principle of social protection and aimed at the conservation of man and nature, as well as at productive organization (Polanyi, 2017, pp. 191–193) and controlled by Nation-States and representative democracy, as will be explained in the next two chapters.

2 The Decline, the Crisis, or the Aging of Capitalist Economy in the Second Modernity

This economic–political system remained balanced for some 100 years. Nevertheless, it contained in itself the origin of the crisis, which mainly had to do with the threat of collapse that hovered above the international economic system and with the impact caused by the gold standard, insofar as the latter actually forced an exacerbation of the national—and later imperial—boundaries that eventually resulted in the two world wars. After all, the gold standard ended up broken, caused an entire transformation of civilization in the 1930s, and triggered the disintegration of world economy as well as of the bourgeois values which had enthroned it. One can consequently state that the origins of this cataclysm can precisely be found in that utopian effort made by economic liberalism to establish a self-regulated market system (Polanyi, 2017, pp. 51–91).

However, an influential role was also played by the type of individualism enthroned by the capitalist system which placed individual existence within a fluid context characterized by permanent change and governed by the increasingly urgent and necessary rhythms of technical innovation. No tradition or habit could thus become consolidated, and neither was any collective universe or stable framework of symbolic references and values—a historical memory—able to guarantee the meaning and expression of true human experience. It cannot be denied that a basic problem of modernity precisely stems from its irrepressible tendency to dismantle every traditional symbolic order, in such a way that crisis and even cultural apocalypse constitute the norm and not the exception. Nonetheless, it is actually the capitalist economy that has intensified this evolution, since its unlimited expansion has everywhere demolished the fabric of habits, traditions, and rites which fixed the limits of each local culture. Neither should we forget that it has exploited resources to such an extent that it has almost exhausted them, wringing everything out of the workforce and tiring it out at a global scale. This has also consequently resulted in the emergence of a widespread feeling of rootlessness among the population, in tune with the growing rhythms of technical innovation, which has established "the absence of the world" (De Carolis, 2017, p. 156).

This explains why the space of flows has replaced the space of places in present-day capitalism (Ingham, 2010, p. 246), why capital has become dematerialized to the extent of having an extraordinary capacity to implement radical innovations in the economy (Bauman & Bordoni, 2016, p. 174) and, in short, why the latter has succeeded in dissociating itself from the society that it manages to dominate (Polanyi, 2017, p. 43). To this must be added that the deepening of globalization has meant the end of the 1945–1990 postwar international order (Amin, 2002, p. 112). Thus, we live within a global economy today, but the political organization of global society has unfortunately proved not to suffice, which means that the development of global economy does not match that of a world society. Furthermore, this type of economy is characterized, together with the free trade of goods and services, by the free movement of capital, which is the most prevalent one. In this respect, capital has a higher mobility level than the other production factors, especially the financial ones, with a flow that exceeds that of direct investment.

No wonder the current capitalist system shows two major deficits respectively related to the shortcomings of the market mechanism and to the deficiencies or the failure of politics. Recession is thus caused by the tension existing between the global dimension of financial markets and the national scope of politics, which remains in the hands of the Nation-State, the basic unit of political and social life (Soros, 1999, pp. 21–158). Or expressed differently, the ultimate source of all contemporary conflicts, and of crises, lies in the tensions between the unyielding principle of *laissez-faire* and the principle of regulation (Mannheim, 2018, p. 32). Since this same tension happens to cause creativity, this means that the latter has finally turned out to be ineffective, that it has failed to successfully solve the typical antinomy of capitalism and representative democracies.

It consequently comes as no surprise that today's capitalism finds itself at a senescence phase in which it is paradoxically fascinated by the new, similar to how old age becomes entranced with youth and the aged capitalism rejuvenates its forms of exploitation. In effect, senility results from several "crises"—linked, among other things, to the accumulation, to the "funding of enterprises, and to the economic and social pauperization [of] a vast majority of the population." The economy is additionally "ill" because it seeks scarcity instead of abundance, just as it happens with medicine, which deals with disease rather than health. In parallel, as the capitalist system grows old, it invests increasingly in technology, facilities, energy, raw materials, and land; that is, more in "dead labor" than in "live labor," the one associated with the work for which individuals are paid a salary. Hence, this uneven proportion represents the real biological indicator of capitalism and simultaneously shows its tragic contradiction. After all, despite needing live labor to endow it with life, capitalism surprisingly discards it, while at the same time becoming updated through new forms of exploitation supplied by the incorporation of new technologies. Nevertheless—and the perverse loop is thus closed—what those technologies make visible is precisely the non-hiring and the marginalization of workers thrown into poverty and precariousness by the forces of capitalism (Amin, 2005, p. 85; Boada, 2017, pp. 63–147).

Therefore, the old capitalism has now becomes a key driver of technological innovation, but not of the social one, which had stood out as the main intellectual motivation of the industrial revolution. As a corollary, capitalist economy has broken away from society and has ended up dominating it (Polanyi, 2017, pp. 43 and 179), thus doing away with the traditionally social nature of capitalism. It "would consequently no longer be the one which opposes capital to labor as a whole, but one which confronts the creditor group (that associates capital and labor) against those who have been excluded" (Amin, 2005, p. 85). This is how the great historical pact between the aforesaid market forces of labor and the democratic ones, which had defined the shaping of representative democracy, has been broken.

3 The Decline of Capitalism, Creativity, and Civilization

This senility of capitalism, this abatement of its initial creative ardor, appears as a clear symptom of the aging experienced by social creative practice in general. As a matter of fact, aged capitalism has become an originator of exclusion, marginalization, and inequality—an issue that will be examined in depth in the next two chapters—rather than the shaper of a new transformative and integrative social utopia. It resembles representative democracy in this sense, since the latter no longer brings society together, but rather expresses how it shatters. However, what characterizes capitalism is that the expulsion of an important population segment, mainly young people, weakens the system and thus simultaneously "ages" it even to a greater extent. This explains, in the opinion of S. Amin, the consolidation of "a permanent chaos which will lead either—at best—to solving it through the beginning of a long transition towards socialism or—in the worst scenario—to catastrophe and to the suicide of mankind" (Amin, 2005, pp. 10 and 155).

Consequently, what the senility of the capitalist system has reduced is the power of instituting creativity, which seems surprising, insofar as the latter was precisely stimulated during times of crisis; such creativity now seems to be shrinking as it had never happened before. It thus deserves to be highlighted that we can hardly find any figure in the intellectual world during the second half of the twentieth century and the early twenty-first century who lives up to those who stood out in the first modernity, to which must be added that culture appears to have lost its bearings in its final moments (Barzun, 2001, pp. 1171–1179). Neither can we forget that the individuality advocated by capitalism and its erosion of all traditions has likewise favored the break of the time narrative inherent to creativity and accordingly severed the close connection existing between the origin and the originating. It will therefore be difficult for capitalism devoid of origin to be able to create something new and to have a future when it lacks a past; in other words, the chances for creativity to be original in the sense examined in this book have become largely reduced.

Most importantly, though, what is really at stake with the weakening of creativity, with the atrophy of its narrativeness, with the decrease of its effervescence, as well as of its emotional and energetic intensity, is the continuity in the construction of human civilization—as the economist mentioned earlier describes it—and even its own survival.

Bibliography

Amin, S. (2002). *El capitalismo en la Era de la Globalización*. Paidós.

Amin, S. (2005). *Más allá del capitalismo senil. Por un siglo XXI no norteamericano*. Editorial Paidós.

Barzun, J. (2001). *Del amanecer a la decadencia. Quinientos años de vida cultural en Occidente (de 1500 a nuestros días)*. Taurus.

Bauman, Z., & Bordoni, C. (2016). *Estado de crisis*. Paidós.

Boada, U. (2017). *La senectud del capitalismo. Un reto a la juventud*. EDlibros.

De Carolis, M. (2017). *La paradoja antropológica. Nichos, micromundos, disociación psíquica*. Editorial Quadrata/Iluminuras.

Florida, R. (2009). *Las ciudades creativas. Por qué donde vives puede ser la decisión más importante de tu vida*. Paidós.

Ingham, G. (2010). *Capitalismo*. Alianza.

Jaspers, K. (2017). *Origen y meta de la historia*. Acantilado.

Mannheim, K. (2018). *El hombre y la sociedad en la época de crisis*. CIS.

Polanyi, K. (2017). *La gran transformación. Los orígenes políticos y económicos de nuestro tiempo*. FCE.

Soros, G. (1999). *Crisis del Capitalismo Global*. La. Sudamericana.

Weber, M. (2017). *Sociología de la religión*. AKAL.

7

Creative Work in the Global Economy of Risk and Uncertainty

1 Introduction

We in the West have precisely had not too favorable an opinion about work. On the contrary, the latter has all through our history and nearly always represented an obligation, when not a torture. However, today, when work is undergoing a deep crisis—as highlighted by the economic and sociological literature—it seems surprising that even so we are unable to revalue it, to succeed in understanding its true meaning for human life, both for individuals and for societies.

Work is certainly in a crisis, having become a scarce good from the past which not everybody can access on equal footing and, besides, the different models set in motion to try to activate it can at most generate only a series of patches that do not serve to solve the root problem. The consequence thereof is that a considerable number of workers above the age of 45 have been unemployed for a long time and will probably not return to the labor market in sufficiently good conditions to maintain their family; that women have more trouble finding a job than men, and that, when they do so, their salaries are significantly lower, thus deepening the historical gender inequality; and also that young people face many

J. A. Roche Cárcel, *Creativity and Time: A Sociological Exploration*, Palgrave Studies in Creativity and Culture, https://doi.org/10.1007/978-3-030-84838-5_7

difficulties in accessing their first occupation, in such a way that their hopes of leading a comfortable life similar to, or better than, their parents' are further and further away.

The enthronement of creativity as an emerging force in the field of Social Sciences, its empowerment as an almost absolute energy, might lead us to think that it could represent an originating force, a useful response to overcome the crisis of work, and even to endow this activity with prestige and to take it to the place where it would really belong. The purpose sought in this chapter therefore consists in confirming whether or not creative work has achieved such aims, if it has really pursued them, and whether or not, in addition to providing quick and abundant profits for creators and for the enterprises that hired them, it has become a powerful driver for social fusion and for the globalization of the economy, of society, and of democracy and its values.

2 The Crisis of Work in Late Modern Society

From the Disregard of Work to the Dilemma: Either more Work or more Free, Creative Time

Work has been poorly regarded throughout Western history (Tezanos, 2001: 11–12). Suffice it to remember in this respect that, when Adam and Eve are expelled from Paradise, as the worst possible punishment, God obliges them to "earn their bread with the sweat of their brow," which means that work is biblically cursed, in sharp contrast with the biblical philosophy of human life, "a present" that we receive and for which we must give something back (Beriain Razquin, 2017: 648). However, it does not seem clear at all that Christianity sees work as being precisely what we must return for the life received as a present, let alone as a gift. Quite the opposite, its derogatory conception will prevail in our civilization rooted in the Judeo-Christian tradition which, following its beliefs, will invent and develop the capitalism where work and the sense that it owns or lacks becomes embedded.

Although manual activity appears as something positive in Ancient Greece, for instance, in the pre-Socratics, in Socrates's thinking, as well as in a few fragments from Plato (e.g. *The Republic*, 535 d.), its overall worldview shows the former as not being worthy of philosophers or of the most educated and highest minds. Instead, it constitutes an obligation of serfs and slaves, whose nature is precisely defined by the need to work. It deserves to be highlighted in this respect that, according to Xenophon, manual trades make those developing them effeminate, and that Plato writes in *The Laws* that a true citizen must not work (Veyne, 1992: 128–130). Aristotle, who links work to matter and to mere life, also despises it by describing work as lacking in freedom—because it prevents leisure—and as something that concerns slaves. Indeed, the *banausia* (manual and mechanical work), along with the slaves who carry it out, arouses rejection in the Stagirite (*Politics*: 1328 a). He additionally regards work as an activity associated with the basic human necessities and as one which transforms nature (Borisonik, 2011: 1–2), but not to dominate it, as will actually happen later on.

The underrating of ordinary—and normally laborious—tasks predominates in Rome too, since they are linked to subsistence and supply, as well to the dependence that they entail. Thus, according to Romans, work turns individuals into slaves rather than into free intelligent beings: "whatever has to do with a salary is sordid and unworthy of a free man, insofar as the salary in such circumstances constitutes the price of work and not of art"—Cicero (Veyne, 1992: 128 ff.). Worse still, the Roman people etymologically connect work with the Latin word "*tripaliare*" and its derivative "*tripalium*," a (torture) yoke for slaves made with three sticks (Naredo, 2002: 38). Consequently, working neither gives us pleasure nor makes us happy; on the contrary, it is a real torment. Without a doubt, the hard working conditions, which demanded a huge physical effort from the body—especially in the countryside, where peasants had to wake up early and endure the cold as well as the heat (Veyne, 1992: 128)—and the poor development of technology justify this conception to a certain extent. Furthermore, people were still hardly aware that work made it possible to transform environmental conditions and, accordingly, to broaden human intelligence, in addition to allowing humans to build an artificial universe of their own which they could inhabit without any risks. Not in vain, without work, these goals become impossible, and so does the development of civilization.

Medieval Christianity's views about work resemble those of Greeks and Romans but, at the same time, it tries to reduce holidays or leisure time—highly abundant in Antiquity—and, particularly in monasteries, attempts are made to venerate labor more and more, thus eventually increasing its appreciation alongside capitalism. This approach will likewise be favored by the search for salvation through work and by other asceticism and mortification practices advocated by some medieval monastic orders and, in the Modern Age, by Luther and Calvin. The nascent capitalism certainly had a positive opinion about work and its "organized" life, the same as monastic and military existence, so that the bells of monasteries and the trumpets of camps and barracks will soon be pleasurably followed by the sirens of factories (Naredo, 2002: 39–40). This revaluation of work continues at the dawn of economic science, as exemplified by William Petty, the seventeenth-century English economist and statesman, who claimed that "the Earth was the mother and Work (was) the father of wealth." This idea gradually took shape with A. Smith—according to whom, a nation's wealth derives from its work (Smith, 2007: 13)—Ricardo and Marx, since Father-Work stopped collaborating with Mother-Earth's productive activities to become the most important component in wealth production, and even the only one, since those thinkers perceived that the Earth could be replaced by Work. The subsequent step in this process took place in the nineteenth century, when Capital, initially a collaborative instrument at the service of Earth and Labor in productive tasks, started replacing them (Naredo, 2002: 42).

This long historical process ultimately led the "political economy"—as it continued to be called by Adam Smith (Smith, 2007: 13)—and the social forces that physically and collectively looked after individuals (the family, the tribe, or the city) to fall into decline or to experience a profound transformation. Then, the productivity-based reason for work collapsed to such an extent that it came to be conceived as an individual and social objective, as well as a means to socialize and to achieve promotion for oneself both professionally and at an economic and social level. Labor was consequently transformed—in the words of Max Weber (1997: 51)—into the key factor within an "intramundane asceticism" regime, in response to the human being's feelings of loneliness and isolation (Naredo, 2002: 43–44). We thus find ourselves at the height of work appreciation.

This does not mean, though, that modern individualism managed to eliminate the criticism of ordinary work or that it succeeded in freeing human beings from their ancestral dominations and dependencies. What modern individualism did achieve was to rationalize the old hierarchization and to preserve it by means of renewed forms, as Veblen points out when he speaks about the "leisure class" and the contempt for ordinary life jobs, typical of the previous hierarchical societies (Veblen, 2008: 39; Figueras & Morero, 2013: 169). The classical Greco-Roman and Christian derogatory vision of work certainly lies at the origin of the modern one, since it remained valid until the nineteenth century and continued to be very active while aristocracy survived. This social class dominated society's general cosmovision and rejected labor as unworthy of their noble condition, and even economic activity and the money stemming from it, even though it allowed them to lead an ostentatious, luxurious life (Sombart, 2009: 70).

Suffice it to say in this respect that a number of important authors have highlighted the negative consequences that the industrial work environment has for social relationships. For instance, Adam Smith believes that, despite generating order, the industrial routine constricts human character and suppresses solidarity, without forgetting his criticism that society progresses in material terms but not morally. Furthermore, factories require little or no thinking or judgment effort from specialized workers (Sennett, 2000: 38–40), who lose autonomy (Gratton, 2012: 21), and become reified—M. Weber refers to the impersonal nature of modern work, to its senselessness, and to its scarce joy (Thaa, 2008: 13)—and alienated (Marx).

Indeed, K. Marx himself, despite his vision of human work as an objectivized activity, which externalizes the essential powers of human beings, and as a production, understood as the generation of something new in the world (Joas, 2013: 148), does not manage to increase the dignity of labor when he inserts it in two concepts: alienation and commoditization. The former is present in *Economic and Philosophic Manuscripts*, where work appears as a key factor in the materialistic and historical conception, insofar as it is the activity mediating between nature and the human being. What is more, it expresses how humans strive to regulate their relationships with nature, building themselves as human beings through its transformation. He additionally understands the alienation in labor, since productive activity within capitalism is lived as a stifling experience that causes suffering, tiredness, and a serious mental as well as physical discomfort. Furthermore, Marx—as it has been traditionally done in the West—opposes work to leisure, the latter being the only one which allows human beings to enjoy themselves and to lead a comfortable life; instead, in the former, nobody hopes to realize their abilities and talents, to develop their creative potential, or to experience pleasure. Labor actually constitutes a sacrifice that humans are forced to "offer" if they want to survive or exist in a more comfortable manner (Fraiman, 2015: 236–237).

As for the commoditization of work, it is analyzed by Marx in *The Capital* and in the context of fetishization of goods. It accordingly comes as no surprise that he regards it as a merchandise similar to sugar with which workers can cover their living expenses and, therefore, survive (Marx, 1985: 9). Consequently, it follows from the Marxian approach that life and work have been dissociated, since the latter is underrated once again by being treated as a value of change rather than one of use. Life may be a gift, but the conditions permitting that life are not given to us humans as a present.

The nearly thirty years that elapsed from 1945 to 1973 witnessed widespread economic progress on the planet, and particularly in the Western world. It certainly seemed that, after World War II, people enjoyed a true "golden age" of economy, which led to reduced hunger and poverty, making it also possible for every country around the world—including the least developed ones—to increase their gross domestic product and to diminish inequalities (Hobsbawm, 2001: 29 ff). Likewise, a Society of Work becomes established on the basis of a social order that fights against poverty and reconnects the individual with society. If we add to this the achievement of full standardized employment, mass consumption, and an economy that relies on security, on certainty, and on the boundaries defined between work, capital, and the democratic State (Beck, 2007: 24 and 99), one can say that this was a period during which democracies enjoyed a gradually increasing stability (Held, 1993: 273).

Who remembered then that labor alienated the worker—as Marx said—in the capitalist production system if everybody was able to buy a fridge, a car, a dwelling, or could afford to take their children to school, or had basic healthcare, an employment insurance, and a decent pension guaranteed for the time when the age of rest and retirement arrived?

We cannot deceive ourselves, though, with the illusion that work leads to progress, that working dignifies human beings, and that it permits to lead a good life, only survived for a few decades. The oil crisis and the essential instability of economy gave way to a real "Age of Uncertainty," which made it impossible to predict a sustained growth in the future and suggested a succession of a whole series of nearly permanent and structural crises instead (Galbraith, 1984); work fell into a deep recession, which has prevented its positive assessment to become fully rooted in the popular imaginary ever since. In fact, "detaching work from the rest of life activities and its subordination to market laws resulted in the annihilation of every organic form of existence and in their replacement by a different type of organization, an atomized and individualistic one" (Polanyi, 2017: 224). For that reason, labor never succeeds in transcending us as human beings, in building us as highly civilized beings, in enabling us to enjoy the post-materialistic democracies announced by Inglehart (1991), which proposed happiness as society's highest value. Its unfair and uneven distribution in the best of cases or the lack thereof in the worst has turned labor into a scarce good—certainly desired but without the aura of the important and significant things in life. That explains why we need to work to build the material conditions of life, but not so much the spiritual ones. It cannot be otherwise because the late capitalist society strongly links work to anxiety, contingency, risk, and uncertainty; work today effectively arises as an "anxiolytic, precarious, unstable, and high-risk activity which not only inflicts physical and physiological wounds but also causes increasingly abysmal social inequalities and a corrosive deterioration of the sense-value of life" (Roche Cárcel, 2013: 3).

In a nutshell, work has not had a good press in the history of Western civilization; working has traditionally been assimilated with a curse or torture, and never been seen as a noble or virtuous activity which builds us as free human beings, and relating it instead to the compelling necessities of society's lowest and most menial classes or segments. How do we conceptualize work at present, when it seems to be scarcer than ever and not equitably distributed? It thus finds itself before a crossroads riddled with uncertainty with which society is very unlikely to cope, since it is contextualized between two irreconcilable extremes. On the one hand, unemployment and compulsive labor, competitiveness, lack of solidarity, and social segmentation jointly and paradoxically increase. It all refers us back to a society which seeks to be democratic and consequently has an egalitarian orientation, but which seems to be imprisoned in a highly mythicized way of understanding labor, while simultaneously proving unable to address the negative consequences that it provokes and to fix limits on the social risks that it entails. On the other hand, there is the desire—or rather utopia, I would say—to consciously reduce the overwhelming prevalence of business production and paid work so that other kinds of more free, creative, and collaborative activities can exist (Naredo, 2002: 47).

The Two Western Work Models

This society additionally questions the prosperity both of economies and of the democratic social liberal State (Held, 1993: 276), and a scenario of total decomposition arises in which labor has also fallen into a deep and perhaps irreparable crisis (Tezanos, 2001: 70). And what should concern us the most is that it looks like a major structural crisis, since it has been caused by the complex interrelation of five forces of change which seem unstoppable and which, together with positive aspects, bring undesired consequences, namely, technology, globalization, demography and longevity, natural resources, and society (Gratton, 2012: 25). Furthermore, the crisis of work is triggered by—and at the same time generates—significant far-reaching structural changes in relation to the traditional industrial society. In fact, it entails the decadence of the specific society and civilization model where it plays an essential role, that of its pattern as employment, that of its connection with salary relationships, and that of its conception as a social activity (Tezanos, 2001: 28–232).

As an example of this profound mutation, the two traditional Western work models (Sennett, 2000: 54) appear to lead citizens to a dead end street. We firstly find the "Rhenish" model typical of the Netherlands, Germany, France, Israel, Italy, Japan, and the Scandinavian countries. The nations where it is applied do not have full employment, but a sort of correction factor exists facilitated by the Welfare State which has reached an agreement to that end with trade unions and enterprises, and which supplies aids to the unemployed, thus mitigating their dramatic situation. However, and despite those efforts, there is no stopping such events as the gradual disappearance of industrial jobs, the increase of early retirements, and the difficult access to employment for young people and women. To which must be added that, pushed by the circumstances or consciously, the countries where this model prevails are making social and labor regulation more flexible to favor entrepreneurs (Tezanos, 2001: 50–51) and are reducing State competences, which implies breaking the old pact between labor and capital, between capitalism, Welfare State, and Democracy (Beck, 2007: 14 and 99; Requejo Coll, 1990: 125). This in turn weakens the safety cushion that the Welfare State used to provide for workers in general and for the most underprivileged in particular. This results in those excluded from the alliance putting the blame on democracy, which consequently loses part of its legitimacy (Beck, 2007: 155).

The second work paradigm—the "Anglo-American" one—is basically implemented in the United States and the United Kingdom. Among its most important characteristics those that stand out are that it has an almost blind faith in free market capitalism and that it conceives state bureaucracy as subordinated to the economy. For its part, the labor market implies a reduced cost, which relates to the insecure, sporadic, and low-paid jobs that it offers (Tezanos, 2001: 50–51). Therefore, greater and deeper social inequalities emerge in this model—between those who have a good, well-paid job and those who are not so lucky—along with increased levels of social violence caused by those who have been left out of the system (Beck, 2007: 11).

It becomes obvious, then, that these two models, the only existing ones until today, place Western democracies before an unsolvable and tragic dilemma: either paying for the growing poverty with a higher unemployment rate, which happens in the first model, or tolerating a scandalous poverty index with less unemployment, which takes place in the Anglo-Saxon American one, albeit reducing it with precarious wages, low productivity, a minimum and insufficient coverage, and greater inequality and crime levels (Beck, 2007: 66). Explaining (which politicians do not usually do, focusing instead on claiming that full-employment policies are feasible) to young people and women, to the long-term unemployed, and to over-45s—the segments with a higher unemployment rate—that these are the only two possibilities available for States to solve their difficulties will definitely not sound too hopeful to them. Both solutions are meager, insufficient, and taking additionally into account that they have already been applied since the 1980s, it would appear that the system as a whole is unable to propose true alternatives, as if it were paralyzed, as if it suffered from lack of creativity, as though it had become stuck, and, in short, as though it had reached the end of a stage.

The Characteristics of Work at Present

In any case, we could delve even deeper into the crudeness of contemporary work by describing some of its key features (Galbraith, 1984: 167 ff; Sennett, 2000: 19–145; Tezanos, 2001: 22–146; Boltanski & Chiapello, 2002: 299–395; Bauman, 2003: 99–175; Innerarity, 2004: 102–108; Beck, 2006: 75 ff; Beck, 2007: 9–165; Beriain, 2008: 142–164 and 204; Gratton, 2012: 21; Roche Cárcel, 2012: 144 ff).

1. The transformation of industrial and labor society into a society of risk is perhaps the most outstanding peculiarity of our time. We may recall in this respect that, according to the definition provided by U. Beck, risk consists in the calculable and quantifiable insecurity that extends to every field—economy, society, and politics—and that its dimensions include globalization, ecology, digitization, individualization, and work politicization. Its most relevant consequences regarding work are vulnerability, discontinuity, imprecision, informality, and, ultimately, the irruption of precariousness. Precarization becomes expressed by means of subcontracting and the growth of temporary and part-time employment, which responds to a twofold strategy of enterprises: a hiring policy which

allows the employer to have freedom and a new scheme for the company's structure which makes it possible for its owners to outsource labor and hide their condition; in this way, they manage to avoid the constrictions of labor law.

Alongside precariousness develops an economy of insecurity and uncertainty too. Both of them portray a dark and blurred future, and, most probably, a dangerous one full of risks. However, what actually represents novelty in all of this is the fact that both the State and the economy pass those risks onto individuals and force wage earners to carry the weight of market uncertainty. If we add to this that the current uncertainty acts as a powerful individualization force which divides instead of uniting, we will become aware that danger is hovering over the everyday life of individuals, filling it with a distressing and uncertain wait, or rather a feeling of risky hopelessness.

2. Work is likewise defined by the despatialization, fragmentation, and acceleration of time. The former is driven by the contradiction existing between global capital and local labor, which acquires a fragmentary nature worldwide. The paradox of this phenomenon lies in the fact that social proximity coexists with geographical distance, or, expressed differently, humans live and work locally isolated and globally connected. Capitalism has thus strengthened the value of the place and increased the desire to form part of a community, but it has simultaneously stimulated the disappearance of barriers, the defrontierization, the deterritorialization, the despatialization of social and political spheres, as well as of production, and the consequent "desocialization" of work, which is leading to its virtualization. As Marx was able to see, when—during the Industrial Revolution—work has the status of a merchandise which can be detached from the person who produces it, it becomes a legal fiction. Nevertheless, now, it has also largely lost its specifically defined location in the factory and has come to occupy an ever-changing space or an imprecise and disorganized non-place where nothing seems to be produced anymore except for concepts or ideas or, what is worse, the abstract benefits detached from the social context, from the capital–labor–State contract. Work is accordingly transformed into something as virtual as the bits which travel at the speed of light inside computers, which once again results in its disconnection from the human body and its disregard as

something mental, fictitious, invisible, and liquid. For this reason, it seems at the very least illogical that, in times of work shortages, and when labor should be better regarded, it does not have more solid features but is contrarily characterized by more ethereal ones, cruelly fitting the prophetic sentence written by Shakespeare, Marx, and Marshall Berman, according to which, in Modernity, "all that is solid vanishes into thin air."

Without a doubt, the deconstruction of work space correlates with a new construction of time which has become disorganized, which has accelerated its speed, and which is represented by a broken, metamorphosed arrow at a point which expresses the maximum and concentrated trust in a narcissistic present. As a cause or a result thereof, capital becomes more "impatient" because of the desire to achieve a fast yield, and thus drives vertiginous and discontinuous institutional changes in the belief that growth will improve. The emphasis is consequently placed on the short term, which makes it difficult for the human being, with a smaller biological nature, to assimilate the speed of the process, which in turn leads—as will be seen in the explanation of the last characteristic dealt with in this chapter—to a "corrosion of the character," particularly of those aspects that bring humans together and grant them the feeling of a sustainable and social self. Indeed, a conflict currently opposes—the lasting—character with a disarticulated experience of time which prevents its consolidation and that of the lasting narrations which define it. This is completed with a flexible time use, especially in large corporations where fixed schedules are less and less common. An example thereof can be found in Toyota, where all idle periods have been eliminated seeking to maximize the working pace, or expressed differently, the level of activity with the same salary. At the same time, there is little appreciation of accumulated experience, stressing the contempt toward the past or the disconnection from it, and concentrating on the immediate instead. As Henry Ford wrote in *The Chicago Tribune* on May 25, 1916, "history is more or less bunk. We don't want *tradition. We want* to live in the present and the only history that is worth a tinker's dam is the history we make today." This passion for the present has gone so far that work currently endows "the ephemeral with duration" and loses its consistency and continuity, memory, and in the absence of the latter, it no longer fuses with the personal biography. Consequently, work becomes a slave to time, and, more specifically, to the present, to the moment which tyrannizes, dememorizes, and freezes our lives at an instant. But, is work also not transformed into a tyrant as a result of this process?

3. Flexibility, deregulation, and informalization emerge as other labor-shaping aspects which stem from an economic power system characterized by three elements: the discontinuous reinvention of institutions, to which I just referred; the centralized concentration of power, which will be explained later; and the flexible specialization of production. The latter has as its aim to achieve increasingly varied and fast products, due to the instability that prevails in consumer demand; and this has to do with the fact that the deregularization of labor relations permits a more rapid transformation of society's institutions.

4. The impact of new technologies along with the passage from an industrial and tertiary society to one of knowledge and information also stand out as important features of work. Four remarks need to be made in this respect. The first one is that the link between the global market and new technologies represents the hallmark of contemporary capitalism, which is no other than the end of labor society, insofar as human beings are replaced by technologies. Indeed, knowledge—not work—currently arises as the main source of social wealth, but the knowledge society does not generate more employment; what is more, it increases productivity without creating jobs. The effects are more than visible since, with the machine-robot, workers become increasingly dispensable, which is added to their alienation, to their social disappropriation, and, in short, to their transformation into social nonsubjects.

The second consideration worth bearing in mind has to do with the so-called technological unemployment which results from a new economic growth model which does not generate any jobs and, when it does create them, they are lower-quality jobs. Therefore, the great paradox about technological societies is that, despite the availability of more wealth, employment is more scarce and worse in them.

The third aspect relates to the global context of employment and unemployment. It is a fact that robotization originates at a faster pace than the new sources of work, and that the latter has not grown since the 1960s—when full employment was achieved—and even remains stable, in the twenty-first century, with an unemployment rate of approximately 30%.

Finally, the fourth observation concerns the impact of new technologies on work and refers to the general consequences derived from the introduction of the Internet: labor decentralization, networked sales with a closer contact between producers and consumers, lower commercialization costs, the organization of work in networks, the easing of contractual regulations, and the spread of new labor patterns, such as teleworking or the subcontracting of self-employed workers, among others.

5. Another factor that shapes labor activity at present is what has come to be known (U. Beck) as its "Brazilianization." Not in vain, we are on the path to replicating the situation of South America, where just about 35% of the economically active population counts on some Social Security protection or help, and where 40% of work is informal. Until now, Europe was supposed to represent the height of civilization, and we assumed that our main goal had to be the universalization of the world, its Westernization. However, precisely the opposite is happening: China and the emergent countries are the economic leaders, and when it comes to work modalities, they are dictated by Latin American countries. As a collateral effect, this situation leads to a loss of self-esteem and of trust in values and in the Western abilities to effectively solve the major labor problems that we are currently faced with—yet another logical consequence of the despatialization of work which entails that of the territory and a deterioration of the solidity associated with its symbols. The desire to establish our values, and very often by force—political messianism—has so far constituted a historical leitmotiv of European and US foreign activity, and it is a close enemy of democracy (Todorov, 2012: 35 ff). In this regard, the fact that Westerners do not see themselves as the best may prove positive to temper their evangelizing ardor, though at a huge cost, since this places them at a clearly disadvantageous situation in economic, labor-related, and civilizing terms, without forgetting the implications and hardships for individuals.

6. Today's work likewise appears as a hierarchized and uneven one. The growth of small and medium-sized enterprises (SMEs) has been paralleled by that of networked organizations, to the detriment of the large enterprises of the industrial age. A similar growth can be attested when it

comes to the influence of large-sized coordinated economic complexes (big corporations, groups, networked organizations of independent firms, and collaborations between enterprises, as well as alliances). And, especially, all markets are witnessing a growing proliferation of oligopolies which broaden their presence and their networks of collaborators beyond their borders. An evolution has thus taken place from organizations with a pyramid-like structure to others shaped by modern institutional networks, lighter at their base, determined—as highlighted earlier—by weaker ties, and defined by networked fragments and nodes; by a structure which no longer has the clarity of a pyramid—it is not simpler, but more intricate—and by a downward, shapeless, and strong domination which becomes more and more evident as human ties weaken. Hence, even though a certain degree of repudiation of authority is theoretically visible among the management staff, the heirs of May 1968, and although they look for new, less classist and hierarchized, and more participatory methods for labor activity, the actual outcome of such efforts is not the disappearance of structural hierarchization in economy, but its adaptation to the new times; to express it in straightforward terms, we are moving from a previous vertical hierarchization to another more horizontal one. In the latter, as usual, the worst brunt is borne by those situated at the—significantly reduced—base, at the lower part, where an increasing number of people are forced to stay at the margin or are pushed to the outside, away from the labor world, and, ultimately, from society and democracy too.

At the same time, a concentration of economic power, without centralization, is taking place as a result of mergers, to the extent that 200 large corporations are under the control of 150 individuals based in five countries. Such a concentration results in an abysmal economic and social distance. In this way, while the most important entrepreneurs are the ones who control, steer, and manage economic affairs—and, more precisely, the macroeconomic ones—the small ones find it increasingly difficult to survive in an environment of giants; and if the rich become richer and richer, the poor are increasingly poor as well. It deserves to be highlighted in this regard that, whereas capital incomes have grown, salaries have decreased in recent years. Power concentration additionally entails an enormous danger, both for the labor universe and for democracy itself. At present, the current strength of trade unions cannot compare to that of those huge corporations, which is partly due to the actual mistakes made by the former but also, and above all, to the successful practices implemented by enterprises to weaken them. Truth be told, mass unionism has evolved until becoming a more professional and less ideologized one, this being a "de-unionization" factor, as attested by the lower membership figures. Otherwise, this weakening of unions has gone hand in hand with a decreased criticism of capitalist enterprises and with a lower number of labor conflicts, which has resulted in a clear depreciation of unions' capacity either to negotiate or to defend workers' rights—the reductions suffered by these rights during the last few years provides sufficient evidence thereof. At a time when economic power prevails over the political one, the existence of a monopoly which concentrates so much power in such few hands does not help at all, to which must be added the smart strategy aimed at the invisibility of this capital, which does not want to be photographed, which has business headquarters in various places, and which hides or makes it difficult to know where this capital pays its taxes or where it uses its profits.

Power concentration is consequently giving rise to a new social hierarchization. This firstly happens owing to the break of the ties existing between capital and labor which had been established during the industrial revolution and were maintained more or less until almost the 1980s and which led to an economic, labor-related, and institutional stability in addition to a growing improvement of workers' labor conditions. Secondly, a dualization of paid work as well as of the labor market is taking place: on the one hand, we have a stable skilled workforce in large firms that benefits from a relatively high salary level and often belongs to unions; on the other hand, there is an unstable, low-skilled, and underpaid workforce that, in addition to having little protection in the small enterprises which deliver ancillary services, is excluded as well. All of this results in a new sort of social polarization between workers—those who have a job and those who do not—which replaces the older one that existed between the aristocracy and the bourgeoisie, or between the latter and the proletariat. This process also affects the middle class, which is being eroded due to the loss of quality in their jobs and the worse salaries that this entails, as well as because of the higher taxes demanded by the financial crisis.

It is additionally worth highlighting that European and US firms have increased both their productivity and their competitiveness, though to the detriment of salaries and doing without the costs generated prior to employment (education, training, etc.) and after it (recovery of forces, and those related to addressing physical wear and aging). Such costs have been assumed by the Welfare State but, after the latter fell into a crisis, it has no longer been able to cover those costs, which has meant the ruin of social capital. For instance, in the United States, voter participation has decreased—only 30% of the population take part in elections—and there has been a "prison miracle" with figures of 500 inmates per 100,000 grown-ups, in contrast with the 80 for every 100,000 adults in Germany. Meanwhile, nearly the same is happening in Europe, with the exception and perhaps only for the time being, of prison and violence data as well as those related to political participation, considerably higher than US figures. Nevertheless, the risk of abstention in the next election calls is very high, which also coincides with the rise of the extremes in the political spectrum, along with that of populist, nationalist, and far-right movements. For these reasons, I consider it appropriate to emphasize that the attacks against the Welfare State and the neoliberal and neoconservative policies which have had such disastrous social effects are being carried out—as Adam Smith would say—by invisible hands, it is true, but not by fleshless or virtual beings; the virtuality of labor has to do with very specific interests and objectives of real economic agents.

7. The last feature which becomes visible in Western contemporary labor activity is its individualization and the "corrosion of character" that is associated with it. First and foremost, it deserves to be highlighted that our time can be described as the age of technical and political transformation of reality, rather than one of eternal spiritual creations. To this must be added that the genesis of the modern technical world shows an inextricable link between natural science, inventiveness, and the organization of work, factors which share rationality as a common element. This explains why work may be defined in three ways: as body work; as a task performed in accordance with a plan; and as the fundamental essence of man as opposed to that of animals, that is, the creation of his own world.

That is why leisure at work depends on the awareness of being involved in the construction of our environment and, therefore, on the worker's ability to become self-conscious in the mirror of what he has produced (Jaspers, 2017: 164–209).

Nonetheless, this has been disrupted with the growth of individuality in labor conditions and, particularly, of wages. Furthermore, this individuality also becomes visible in the deteriorating conditions of teamwork, since the latter has acquired a more superficial nature, in such a way that the—labor—community has turned into a "fiction." Likewise, the disintegration of the work community comes as a result of using, in the same space, workers from diverse firms and subject to different labor regulations, which leads to dismantlement and disorientation in collective action, and, in short, to an individualization of tasks.

However, it is not a true substantial individuality based on singular ideas and emotions that we are speaking about here, but rather a type of individualism which resembles that of the lonely activity of consumption, and which is accordingly more consumed than lived. The worker's private and emotional life goes adrift before the growing difficulty to reconcile personal and family responsibilities with those related to work. It comes as no surprise that, gradually and without any limits, workers are being forced to accept more flexible working hours, full availability, and a level of responsibility which increases gradually but steadily, to which must be added that they are under huge pressure to reach specific objectives and goals, to compete in an extremely cruel and unbearable world where profits must be more and more abundant and achieved increasingly fast. Thus, work of course no longer personifies pleasure, but rather a *tripalium*, an instrument of torture: we have made such little progress from the origin of the concept to the present day! Furthermore, it no longer provides an ethical model, insofar as a conflict arises between its values—a short-term purpose, an instrumental rationality that depends on goals rather than on emotions, competitiveness, lack of solidarity, and the personal or family-related ones—for example, duties, honest behaviors, solidarity, or empathy and understanding toward others.

This explains why the worker's character has been corroded little by little, why labor no longer constitutes a fundamental source to shape personality, but quite the opposite. After all—as highlighted earlier—if character builds upon the lasting, "long-term" aspects of our emotional experience, a short-term task which neglects the past, which does not take experience into account, which leaves emotions aside, and which replaces human intelligence by a mechanical one, does not favor the formation of character anymore. Due to all of these, work is not legible for the worker either, since the latter does not understand what he is doing or, in the best of cases, he only does so in a superficial way, which pushes workers toward indifference and a fragile labor identity.

3 Creative Work in the New Global Economy

On the Concepts of Creativity and Global Economy

This overall context of work crisis develops in the current global economy, which, determined by the New Information and Communication Technologies—social media and the Internet in particular—has stimulated and pampered creative activity. Nonetheless, what I will try to unravel here is if the latter manages to diminish or temper the situation of work crisis and whether or not it succeeds in improving such a precarious situation. With that aim, I will make an effort to describe the function performed by creativity in global economy, additionally providing some brushstrokes to show what I understand by global economy and creativity.

The economy of globalization shows high geographical dispersion and mobility and simultaneously a territorial concentration of resources for service management and delivery purposes. Therefore, the more globalized enterprises become, the more their central management functions grow (Sassen, 2007: 80 ff.) and the more they help the cities in which they are installed to be labeled as "global." Nonetheless, for a metropolis to reach the consideration as global, it needs to have financial centers, the presence of international corporate head offices, the development of business and commercial management services, a well-developed transport hub, international tourism, a considerable population size, a multicultural mix of national and foreign inhabitants, and a concentration of artistic and scientific elites (García-Canclini, 2005: 167).

In turn, creativity can be described as an innovative or valuable idea or action, or "the formulation of new ideas and their implementation to elaborate original works of art and cultural new ideas and cultural products, functional creations, scientific inventions and techno-cultural innovations" (Boix & Lazzeretti, 2012: 182–183). Needless to say, innovation—that is, the translation of technology into inventions—has become the basic component to compete successfully in the economy of globalization, marked by the New Information and Communication Technologies. This innovation has two essential features, though: immateriality and a democratic nature (Anuri, 2001: 121–125). In effect, "the economy of access" (Rifkin, 2000: 50)—which can be defined as cultural, virtual, networked, or access-based, and relying on concepts or ideas—has replaced the capitalism of industrial production and of ownership. This is very clearly exemplified by the Nike sports footwear brand, originally designed with an insole or model which follows a certain idea of design with specific materials, colors, and prices that the subcontracted (usually Asian) factories must faithfully stick to. Consequently, instead of manufacturing trainers, Nike sells an idea of them.

Creative Cities and Creative Industries Cause New Social Inequalities

Therefore, the globalization of the economy and creativity go together, insofar as the former has made possible the emergence of creative cities. These are supported on the "geography of happiness," since regions, the same as individuals, have well-defined personalities, which is of paramount importance to understand them, alongside the phenomenon of migrations and economic growth. Thus, geographical location continues to stand out as a determining factor in global economy and its key resides in the interaction of personality and place (Florida, 2009: 74–285).

However, as could be imagined, the economic growth of these major cities has not been driven by specialization, but rather by innovation and flexibility—the central element in creative economy—together with geographical location, an essential factor in global economy. It is consequently this relevance of the place, linked to its personality, that brings wealth. Here, in these creative cities, their nucleus comprises a "new class of scientists, engineers, university lecturers, poets, actors, novelists, animators, artists, architects and designers, cultural workers, scientific thinkers, analysts and opinion dealers whose economic function consists in developing new ideas, new technologies, and/or a new creative content" (Florida, 2010: 8).

These creative cities host a booming creative culture- or awareness-related industry (Boix & Lazzeretti, 2012: 196) and a "creative class" to which capitalism has entrusted its own aesthetic reproduction in the new technological age (Delgado, 2016: 73). That is the reason why, unlike "cultural industries," the "creative" ones have lost their cultural and democratic sense, as "leisure is directed and entertainment programmed" in them (Muñoz, 2015: 222), and because they have been exclusively oriented toward the market context and that of short-term profitability. That helps us understand why creativity has become a concept overloaded with economism, the result of which has been the displacement of culture in favor of creativity. This development acquires more importance in the present times of crisis, since creativity has been confused with innovation and has stressed the economistic bias of politicians, highlighting technological determination, placing the emphasis on the insensitiveness toward culture, and highlighting the competitive vision of globalization, which pursues disintegration rather than fusion. In addition to this, insisting on economism instead of culture entails an irreparable democratic loss, insofar as culture serves as the vital cement, the main value of democracy. For that reason, in contrast with this profitable short-term culture, support should be given to a cultural training per se as well as to cultural policies oriented to favoring social creativity from the base (Bustamante, 1995: 18–23).

Digital creative industries do not seem to be keen on this, though, or, in any case, we could acknowledge that they are strongly influenced by an ambivalent as well as paradoxical panorama (Rodríguez Ferrándiz, 2011: 151–155). After all, these industries are causing the overlapping of leisure and work times and spaces, the industrialization, rationalization, and commercialization of the former, and an informal look at the latter.

Alongside this, the conversion of the typical cultural product—the book, the record, the film, or the videotape—into a digital file or a document which can be accessed online, and its assimilation with a device like the computer, leads, as in the New Technology industries, to such a radical dematerialization, "transfiguration," or "transubstantiation" that it prevents us from labeling those new figures or substances as "matter." Hence, this transfiguration of cultural goods encourages a type of creativity which is simultaneously democratic and nondemocratic and, therefore, entails an active and contemplative creative "experimentation."

In addition to this, the democratic side of creative economy in global society is more theoretical than real since, in practice, instead of creating equality, it actually generates a deep and growing inequality between the different nations, as well as between individuals, marginalized and defined by a new form of illiteracy—in this case, associated with the scarce production, knowledge, and use of technologies (Anuri, 2001: 121–125)—or hugely enriched thanks to their creative talent. In relation to this, Antonio Ariño, in *La Secesión de los ricos* [The Secession of the Rich] (2016: 62–94) precisely insists on the fact that the creativity of technological and financial societies causes new inequalities, which explains why late modern possessive individualism borrows the ideas of personal autonomy, merit, and talent from the democratic movements of the 1960s (Boltanski & Chiapello, 2002: 241 ff) and, accordingly, why it legitimizes the scandalous profits of a business elite and underpins inequality.

Thus, as added by Ariño (2016: 66–127), within the four current categories of rich people—geeks, sheikhs, oligarchs, and bankers—the former are geniuses in specific fields of engineering and computational science who achieve their status precisely because of their initiative, inventiveness, and boldness. In turn, the high salaries and gains of financiers have been transformed into the incentives needed to take chances and innovate, as well as to almost endlessly increase their personal wealth. To this must be added that the 2008 crisis has led the ultrarich to undertake a geographical diversification of their investments, to devise new tax evasion strategies, thus building an ecosystem of mobile money which promotes huge technological, social, and cultural innovations. Among them stand out offshore financial centers and spaces for technological innovation and invention. As opposed to these privileged locations, in the peripheries of the capitalist system, poverty and the uneven distribution of wealth derive from the actual polarizing logic of global economy (Amin, 2002: 31).

All in all, if the most decisive and most commonly debated-upon issue when it comes to democracy is probably the relationship between equality and freedom, and between the latter and power (Flores D'Arcais, 2013: 83), creativity does not appear to have had a positive influence on such links. On the contrary, it has fractured them, contributing to increasing the distance between democracy and capitalism.

This allows us to assert that, if the creativity of global economy has opened an abyss between capitalism and democracy, the creativity of cities and creative industries has done the same thing between economism on one side, and culture and democracy on the other.

4 An Unequalizing Creative Work

As verified in this chapter, the contemporary labor crisis results from the deep polarization and inequality stimulated by global economy, which, determined by the New Information and Communication Technologies, has encouraged and pampered creative activity. However, the latter has been unable to alleviate or temper the situation of crisis in which work is immersed and has failed to improve its precarious situation. The current creative work has certainly not succeeded in reducing unemployment rates and has not even managed to dignify the status of labor activity to a substantial extent. As a matter of fact, we could be led to believe that creativity—so fashionable during the last two decades of the twentieth century and the beginning of this twenty-first century—could mean a significant change in the enthronization of work as an essential activity for the human being, meaningful both at an individual level and socially. Nonetheless, it has been attested that creative work has not become the manna that miraculously rains over the thousands of unemployed people across the globe. Quite the opposite, it has brought about new inequalities, showing the structural defects of a system, the capitalist one, which in addition to being profoundly unfair by nature, can no longer offer an occupation—quantitatively and qualitatively—to everyone who wants one; what is more, it seemingly has no interest in doing so because it pays more attention to, and focuses to a greater extent on, collecting larger and larger profits in gradually shorter and faster times.

On the other hand, evidence has been supplied here that creative work is characterized by at least five essential aspects:

1. It is based—the same as work in general—on risk, flexibility, and inequality.
2. It constitutes a factor of competitive individual development and, as a corollary, of social fracture.
3. It is not helping remove the bad reputation associated with the traditional concept of work, without increasing the availability of spare time either. In any case, what it definitely does instead is—exactly like contemporary capitalism—to fuse business time with time dedicated to leisure, to merge working hours with party time.
4. It relies to a greater extent on innovation than on creativity itself, and it does not lay emphasis on the origin (the past) but on novelty (the future), because it focuses on structural transformation instead of the short term. It has thus broken the time narrative inherent in the creativity which I have been referring to throughout this book.
5. The creativity that it displays has been fundamentally supported on social change, though the focus on the most profitable innovations has commoditized this change and mixed it with the rhythm of goods. Hence its lack of interest in social, reformist, or revolutionary transformation, and the greater attention paid to the leading industrial sectors, making them more powerful. This ultimately explains why it has mutated into yet another merchandise—as Marx saw—and into an unequalizing instrument at the service of the global capitalist system.

Therefore, creative economy has widened the gap between the market and the State, between politics and economy, between capitalism and representative democracy, even activating the contradictions that characterize capitalism. In other words, it has deepened the crisis, the decadence and collapse of democracy, in addition to generating new inequalities between the rich and the poor and between creative and noncreative States.

Lastly, I think we have been able to confirm here that the creative global economic system has commoditized creativity, depriving it of meaning, and has empowered it against culture, thus contributing to worsen the crisis of representative democracies and of their values and, most importantly regarding our interest in this chapter, to deepen the crisis of work itself.

Bibliography

Amin, S. (2002). *El capitalismo en la Era de la Globalización*. Paidós.

Anuri, D. (2001, November). La Democracia reforzada por la creatividad. *NUNTIUM*, 119–124.

Ariño, A. (2016). *La secesión de los ricos*. Galaxia Gutenberg.

Bauman, Z. (2003). *Modernidad líquida*. FCE.

Beck, U. (2006). *La sociedad del riesgo global*. Siglo XXI.

Beck, U. (2007). *Un nuevo mundo feliz. La precariedad del trabajo en la era de la globalización*. Paidós.

Beriain, J. (2008). *Aceleración y tiranía del presente. La metamorfosis en las estructuras temporales de la modernidad*. Anthropos.

Beriain Razquin, J. (2017). La metamorfosis del don: ofrenda, sacrificio, gracia, substituto técnico de Dios y vida regalada. *Política y Sociedad, 54*(3), 645–667.

Boix, R., & Lazzeretti, L. (2012). Las industrias creativas en España: una panorámica. *Investigaciones Regionales, 22*, 181–206.

Boltanski, L., & Chiapello, É. (2002). *El nuevo espíritu del capitalismo*. Akal.

Borisonik, H. (2011). Pensando el trabajo a través de Aristóteles. *Astrolabio. Revista internacional de filosofía, 12*, 1–8.

Bustamante, E. (1995). De las industrias culturales al entretenimiento. La creatividad, la innovación…viejos y nuevos señuelos para la investigación de la cultura. *Diálogos. Revista Académica de la Federación Latinoamericana de Comunicación*.

Delgado, M. (2016). *Ciudadanismo. La reforma ética y estética del Capitalismo*. Catarata.

Figueras, A. J., & Morero, H. A. (2013). La teoría del consumo y de los ciclos en Thorstein Veblen. *Revista de Economía Institucional, 15*(28), 159–182.

Flores D'Arcais, P. (2013). *¡Democracia! Libertad privada y libertad rebelde*. Galaxia Gutenberg-Círculo de Lectores.

Florida, R. (2009). *Las ciudades creativas. Por qué donde vives puede ser la decisión más importante de tu vida*. Paidós.

Florida, R. (2010). *La Clase Creativa. La transformación de la cultura del trabajo y el ocio en el siglo XXI*. Paidós.

Fraiman, J. A. (2015). Algunas consideraciones sobre el concepto de trabajo en Karl Marx y el análisis crítico de Jürgen Habermas. *Trabajo y Sociedad*, No. 25.

Galbraith, J. K. (1984). *La Era de la Incertidumbre*. Plaza & Janés.

García-Canclini, N. (2005). *La globalización imaginada*. Paidós.

Gratton, L. (2012). *Prepárate: el futuro del trabajo ya está aquí*. Galaxia Gutenberg-Círculo de Lectores.

Held, D. (1993). *Modelos de democracia*. Alianza.

Hobsbawm, E. (2001). *Historia del Siglo XX*. Crítica.

Inglehart, R. (1991). *El cambio cultural en las sociedades industriales avanzadas*. CIS.

Innerarity, D. (2004). *La sociedad invisible*. Espasa Calpe.

Jaspers, K. (2017). *Origen y meta de la historia*. Acantilado.

Joas, H. (2013). *La creatividad de la acción*. CIS.

Marx, K. (1985). *Trabajo asalariado y capital*. Planeta-Agostini.

Muñoz, B. (2015). *La dominación simbólica en la globalización. Una Teoría Crítica sobre la Postmodernidad*. Fundamentos.

Naredo, J. M. (2002). Configuración y crisis del mito del trabajo. *Scripta Nova: Revista electrónica de geografía y ciencias sociales*, Extra Issue No. 6, 119.

Polanyi, K. (2017). *La gran transformación. Los orígenes políticos y económicos de nuestro tiempo*. FCE.

Requejo Coll, F. (1990). *Las democracias. Democracia antigua, democracia liberal y Estado de Bienestar*. Ariel.

Rifkin, J. (2000). *La era del acceso. La revolución de la nueva economía*. Paidós.

Roche Cárcel, J. A. (2012). Tiempo líquido y cultura de la incertidumbre. *International Review of Sociology, 22*(1), 137–162.

Roche Cárcel, J. A. (2013). El incierto sentido del trabajo en la sociedad de riesgo global. Barataria: revista castellano-manchega de ciencias sociales, ISSN 1575-0825, No. 15, 207–226.

Rodríguez Ferrándiz, R. (2011). De industrias culturales a industrias del ocio y creativas: los límites del "campo" cultural. *Revista Comunicar*, 36, *La televisión y sus nuevas expresiones, 18*, 149–156.

Sassen, S. (2007). *Una sociología de la globalización*. Katz.

Sennett, R. (2000). *La corrosión del carácter. Las consecuencias personales del trabajo en el nuevo capitalismo*. Anagrama.

Smith, A. (2007). *La riqueza de las naciones*. Alianza editorial.
Sombart, W. (2009). *Lujo y capitalismo. Economía de la ostentación*. Sequitur.
Tezanos, J. F. (2001). *El trabajo perdido. ¿Hacia una civilización postlaboral?* Biblioteca Nueva.
Thaa, W. (2008). Democracia y crítica de la civilización en Max Weber y Hannah Arendt. *Revista española de ciencia política, 19*(October), 9–40.
Todorov, T. (2012). *Los enemigos íntimos de la democracia*. Galaxia Gutenberg-Círculo de Lectores.
Veblen, T. (2008). *Teoría de la clase ociosa*. Alianza.
Veyne, P. (1992). El Imperio Romano: Los esclavos; Trabajo y descanso. In P. Ariès & G. Duby (dirs.), *Historia de la vida privada. Imperio Romano y Antigüedad tardía* (Vol. 1). Taurus.
Weber, M. (1997). *Sociología de la religión*. Fondo de Cultura Económica.

8

The Connection Between Creative Economy and the Crisis of Representative Democracy

1 Introduction

Hans Joas's Theory of Creative Action and the Way Out of the Crisis

For this reason, the present work follows in the footsteps of the books *The Creativity of Action* and *Creativity, Action, and Values* authored by the sociologist Hans Joas where he argues that active creation can offer a way out of cultural crises, especially the one suffered by representative democracy. Joas (2013, pp. 62 ff. and 309; 2002, pp. 123 and 131–132) has tried to lay claim to the concept of creativity and to leave behind the marginalized status traditionally assigned to it by Sociology, despite several important classical antecedents—G. Simmel, E. Durkheim, F. Tönnies, M. Weber, or C. Marx. And he does so by depicting its most characteristic features, since it gives sense to action and liberates new actions, it is intellectual and emotional, physical and individual as well as collective; in this regard, Joas considers that, if society is the product of creative activity, so is the individual. He additionally understands—following John Dewey—that different degrees of creativity exist and is aware

J. A. Roche Cárcel, *Creativity and Time: A Sociological Exploration*, Palgrave Studies in Creativity and Culture, https://doi.org/10.1007/978-3-030-84838-5_8

that the latter transforms values into facts, into elements which allow for a decent existence, or, expressed differently, which "make the world" (Sánchez Capdequí, 2014, p. 211).

Furthermore, Joas reminds us that it is no longer possible to have a rational and normative kind of society today. In fact, he links creativity both to democracy—"creative democracy"—(Joas, 2013, p. 253 and ff.) and to the period of crisis that we are going through. Not in vain, the crisis of modernity is associated with the crisis of representative democracy, and no wonder its concept is mobile and fluctuating, since it evolves jointly with society (Bauman & Bordoni, 2016, pp. 139–178). In this respect—according to Bauman—late modernity breaks up with modern normativity and rationality, giving rise to an uncertain, contingent, and risk-based society where all our actions have unexpected consequences. Therefore, along these lines continues the anti-theological and anti-evolutionist approach to creative action advocated by G. H. Mead, in whose opinion democracy constitutes an institutionalized revolution, and for whom social change is not rationalist but permits the new to emerge constantly and unpredictably; it is accordingly contingent. This consequently reveals both an increase of individual options for action and a greater awareness of the actual contingency of History.

Nevertheless, as regards Joas's creativity of action, it is worth highlighting other important aspects, insofar as that would explain the autonomy of the agent as well as social change (Joas, 2013, p. 15; introduction by Ignacio Sánchez de la Yncera), and because his theory prioritizes the empirical and social contexts where creative action takes place (Farfán, 1999, p. 56), without forgetting that he defines our behavior as persistently specific, concrete, individualized, and unique. This justifies why Joas does not aspire to build universal action models for action, but rather to carry out the "historical reconstruction of social practices." In this respect, it makes a lot of sense to state that "creative action would thus be the psychological and social ability that any actor owns to respond innovatively when faced with the problems or challenges which arise in a specific situation and which lead him to revise the routines and habits in accordance with which he has behaved on a daily basis" (Farfán, 1999, pp. 56–61). In the end, this shows that the conditions created by situations are not limiting; they work as facilitators instead (Joas, 2002, p. 16; introduction by Gustavo Leyva and Oliver Kozlarek).

Therefore, Joas's theory of creative action is not only innovative but also rescues an essential topic within sociological theory that deserves to be developed: the notion of creativity, directly linked to the theory of action and to that of social change (Cristiano, 2010, p. 1 ff.; 2012, p. 33 ff.). Furthermore, the proposal made by Joas enables Sociology to continue enquiring about creativity and to look for research alternatives, as I will try to do in this chapter.

More specifically, it is my intention to complete his original contribution in various senses, along the lines of what I have been suggesting in this book in relation to the real meaning of creativity, for which purpose special attention will be paid to the concepts of contingency and indetermination. This issue has also aroused the interest of the US sociologist Immanuel Wallerstein (1997, pp. 69–71), who, based on the reflections about their interrelations, has pointed out that we currently find ourselves in a transition period, full of turbulence, with very high risks and uncertain results; and also of some Spaniards (Bergua et al., 2016, p. 2) according to whom creativity questions any established order, insofar as it consists in causing something unforeseeable to appear, which in turn relates it to risk, which—like creativity—has a contingent nature too.

In fact, from my standpoint, precisely that contextual indetermination—added to the one typically linked to creativity—does not permit to categorically state that it can prove useful to solve the crisis of representative democracies and leads me to insist not only on its possibilities but also on its limits, to stress both its negative aspects and the positive ones. And, although creativity is basically associated with social change, I will likewise try to relate it to the structure of global contemporary capitalism. Finally, I will focus, albeit succinctly, on establishing some connection between creativity, social interrelations, and power.

Around the Concept of Crisis

The word *Krisis* derives from the Greek verb *krínein*, which means "to examine," "to decide" (Thom, 1979, p. 69). However, after becoming widespread at present, the notion of crisis has undergone a sort of conceptual emptying; whereas it did originally mean "decision," that is, "the decisive moment," today it expresses indecision or relates to uncertainties (Morin, 1979, p. 277). In any case, a crisis could be defined as a major turbulence or disruption of the social system triggered by external or internal causes (Thom, 1979, pp. 70–71) that is considered beyond its duration and geographical scope, and which can also endanger the actual existence of that social system and its essential reproduction mechanisms. Nevertheless, it can also be recomposed on some occasions, rejecting what is harmful and incorporating innovations meant to rectify the situation. In the first case, crisis leads to decay, and subsequently to collapse, whereas in the second, it makes possible a more or less effective and lasting recovery (Beinstein, 2005). Therefore, either the crisis will degenerate into a catastrophe or it will end up being reabsorbed (Thom, 1979, p. 69), and it is for that reason that it can be observed not only as a simple balance modification but also as the passage from one order to another (Attali, 1979, p. 165).

As a matter of fact, crisis itself represents the change from a state regarded as normal to another evolutionary and transient one (Thom, 1979, p. 69). Indeed, a crisis is a "moment of accelerated mutations in the conditions for the reproduction of social organizations" (Attali, 1979, p. 181), a transition stage, an intermediate space–time between continuity and revolution or between permanence and change (Starn, 1979, pp. 12 and 28). And that is precisely why the study of crises becomes essential for a true exercise of sociological imagination, since the overall problem of History in this connection needs to be interwoven with those affecting social structure or, expressed differently, it must interrelate order and change (Mills, 1999, pp. 65–66).

From a sociological point of view, and more strictly speaking, a crisis arguably represents a collective situation characterized by a series of contradictions and ruptures which is additionally riddled with conflicts, tensions, and disagreements (Freund, 1979, p. 190). Furthermore, it can contribute or not to social development, as explained by Saint-Simon—who exerted a remarkable influence on Auguste Comte and E. Durckheim in relation to these matters—for whom the evolution of mankind can be analyzed as a succession which alternates "critical" and "organic" "periods" or "states." It is precisely during this critical stage that every communion of thinking, of joint action, and of coordination comes to an end, and that beliefs are abolished. Likewise, the uncertainty of general relationships is transmitted to private ones, consequently altering "exchanges of meaning" with "economic exchanges," that is, the privileged means to assure the cohesion of social systems. Therefore, the overall social crisis can be analyzed as a breakdown in the legitimacy associated with the system of values (Béjin, 1979, pp. 114–115).

The Break Between the Market and the State

Globalization is characterized by an economic integration based on replacing relationships by transactions, since it allows for freedom in the trade of goods and services as well as the free movement of capital, information, and entrepreneurship, in addition to a close interrelation between interest rates, exchange rates, and stock prices in various countries. This connection—and the huge benefits that it entails—perhaps explains why the global capitalist system has such a favorable attitude toward financial capital and why global financial markets currently exert a decisive influence on the overall economic conditions (Soros, 1999, pp. 21–159).

This global capitalism no longer relies on producing money through goods, though; instead, it does so by means of knowledge (Fumagalli, 2010, p. 89) and creativity, turned into economy's most important "asset." The height of human creativity has become the prototypical mark of economic life and "the force which drives the major changes in our era" (Florida, 2010, pp. 44, 45 and 61), to which must be added that creativity is conceived as a new civilizing *telos*, the great objective that can facilitate development and the exit from the structural crisis of our time. No wonder it is in fashion, so much so that even Sociology pays attention to it and everyone aspires to consuming it, to buying it for a price, and to demanding that creativity as if it were a right (Pope, 2005, pp. 39, 40), which, combined with the possibility of choosing that consumers have and their "democratic" freedom—confused with the one that defines global economy—has paradoxically endowed creativity itself with power and superficiality. In other words, nowadays creativity is less substantial but also more omnipresent, multidimensional, and multifaceted, while at the same time constantly interacting with other communication-related, technological, artistic, and cultural components, and being framed within a social and economic environment that feeds it (Serrano, 2014, p. 37) and which is nourished by it as well. In fact, today we are witnesses of an extensive publishing activity in the field of academic literature dedicated to "creative economy," to "creative classes," to "creative cities," to "cultural industries," and to "creative industries," which produce goods adapted to mass consumption (Adorno, 1991, p. 98).

Consequently, that global economic system, its producing agents and the clientelist masses—as will be explained in this chapter—have commercialized creativity and stripped it of its substance, rendering it meaningless, and thus contributing to deepen the crisis of representative democracies.

This is, above all, because tension has been generated between the global context of financial markets and the local sphere of politics, whose basic unit continues to be the Nation-State. After all, even though governments keep the power to intervene in the economy, they are increasingly subordinated to the dictates of worldwide competition for trade and capital. Therefore, capitalism has come to be confronted with representative democracies, and as a result of this fight, global economy development does not coincide with that of global society, which has led to the failure of politics and the discrediting of the Marxist, liberal, and social-democratic ideologies, as well as of reformism and revolution (De Souza, n.d.), of the Welfare State (Amin, 2002, p. 51), of traditional political parties, and of democratic institutions, which have entered into a deep crisis of values and legitimization.

Economic globalization has accordingly given rise to a world which is still fragmented in different nations and divided between the rich and the poor, as well as to a revival of nationalisms and social fights, together with the rise of ethnicity, a political response to economic globalization (Amin, 2002, pp. 38, 57 and 75–95). In a nutshell, it is now proven that the management of global capitalism has broken the demand "for the market in addition to the State," something which had been one of the hallmarks of capitalism, accustomed to the historical coexistence and the mutual dependence of two types of power: the economic private one stemming from ownership and profits; and the coercive territorial power of States (Ingham, 2010, p. 211).

This chapter will precisely deal with the role played by creative economy in this context of crisis, especially focusing on ascertaining whether or not it has favored the break between market and State, or, expressed differently, whether or not it has worsened the contradictions of capitalism itself, and also if it has made the latter unable to overcome the tension existing between economic management, growingly globalized and creative on one side, and political and social management, where creativity still does not seem to have arrived, on the other.

2 About the "Crisis" of "Representative Democracy"

What Is Representative Democracy?

It is no easy task to define democracy, since at least four different perspectives arise on this continuum: constitutional; substantive; procedural; and process-oriented (Tilly, 2010, p. 37). However, rather than simply making another characterization, what I am going to do here is to draw a conceptual map which makes it possible to specify as accurately as possible what we are talking about when we refer to it. In this sense, it seems to me that it would prove highly useful to compare the features that define classical democracy, the original one, and the representative one, its heiress, for the purpose of identifying not only the differences but also the similarities between them. Thus, I will take as a starting point Harold Valencia (2016, p. 125 ff.) and the features that he establishes in reference to present-day democracy and the Hellenic one, albeit qualifying and questioning his emphasis on the differences and completing the information that he provides as much as possible:

1. In Greece, the constituent and institutive power was vested on citizens (Marangudakis, 2016, p. 774 ff.), who personally and directly participated in the council, the courts, and the Assembly (Roche Cárcel,

2013, p. 25); instead, present-day power is exercised by means of political representation. Nevertheless, participation in Antiquity was limited and not as direct as we usually believe (Marangudakis, 2016, p. 774 ff.)—in sharp contrast with what happens today—as it is restricted, even though citizenship has become almost universalized; the integration of immigrants remains a pending issue and, furthermore, migratory phenomena have shrunk the concept of citizenship (Arango, 2003, p. 19).

2. In Greek democracy, the public sphere is a concern for citizens, as opposed to representative democracy, where public affairs are left in the hands of the representative elites (Beetham, 1985, pp. 5–6) or of specialists, which has enlarged the distance between the culture of experts and that of the general public (Habermas, 1984, pp. 27–28; 1989, pp. 63–67).
3. In the ancient world, there is no constitution, or constitutional document, which actually serves as the fundamental referent for representative democracies—with the exception of the United Kingdom.
4. In Greece, the polis is made up of three institutions—a central government; an oligarchic council; and a general assembly of citizens—which complement one another, collaborate, and offset power (Roche Cárcel, 2013, p. 26). At present, the division of peers has become systematized, which guarantees both a balance and a monitoring of public powers (Montesquieu, 1997; Arendt, 1988, p. 155; Baños, 2006, p. 39), although the executive power usually prevails over the others.
5. In the ancient world, democracy was associated with mortality and limitation; that is to say, democracy implied thinking of limits. For this reason, chaos underlay in any political issue and finitude arose as the paradigm of both the human sphere and the cosmos (Vernant, 2003, p. 32 ff.; Morin, 1981, p. 76 ff.; Castoriadis, 2006, pp. 10 and 45–358). Well, that culture of limits led democracy to rest upon a religious foundation (Rodríguez Adrados, 1983, p. 21 ff.) and to have a link with freedom (Castoriadis, 1998, p. 114 ff.; 2008, p. 156) and with risk (Nussbaum, 1995, pp. 30–42). Instead, today's culture rejects any idea of mortality, since death is concealed (Ariès & Armiño, 1987, p. 95 ff.) in addition to being individual (Morin, 1994, p. 40

ff.) and lonely (Elias, 2012, p. 19 ff.). It comes as no surprise to check that hardly anybody reflects on the limits of democracy, save for Machiavelli, Montesquieu, R. Dahl (Roche Cárcel, 2013, pp. 51–52), and H. Arendt (1999, p. 46). In any case, contemporary democracy is atheistic by nature, and accordingly opposed to theocracy (Flores D'Arcais, 2013, pp. 16 and 58).

6. *Paideia*, the education of the individual, was key for Athenian democracy since, according to it, the citizen is not born as such, but becomes one in the process of socialization, of humanization; moreover, the individual cannot exist without tradition and the community (Rodríguez Adrados, 1983, pp. 74, 110 and 441). At present, prevalence is given to a society of hedonistic (Bell, 1992, pp. 11–63) individualization (Beck & Beck-Gernsheim, 2016, p. 37) where the citizen has gotten rid of the community and in which the public sphere has been colonized by the private one, since "public affairs" have become the private problems of public figures and, therefore, the public context has been transformed into the space dedicated to the public confession of private secrets and intimacies (Bauman, 2003, pp. 13–86; Flores D'Arcais, 2013, pp. 110–111). To this must be added the process of growing depoliticization that universities are experiencing, since they depend more and more on economic demands. We consequently find ourselves further from a "citizen university," from a real "university democratic matrix" (Beltrán Llavador, 2014, pp. 5–6).
7. The dominant political *Ethos* among Greeks was their sincerity, since they considered that law only exists between equals (Castoriadis, 2008, p. 156). Nowadays, we say one thing and do a completely different thing, without forgetting that any truth has been devoured (Vattimo, 1995, pp. 24–159), replaced by an insecure existence lacking in certainties where politicians resort to lying as a weapon to beat their enemies and "people swallow all the lies" (Fernández-Armesto, 1999, pp. 177–215 ff.; Arendt, 1999, pp. 9–57). To make matters worse, the border between information, entertainment, and fiction is becoming increasingly blurred. Consequently, lying in politics has become a usurpation of sovereignty, insofar as the truth, the whole

truth, and nothing but the truth should be the categorical imperative of democracy (Flores D'Arcais, 2013, pp. 46 and 51).

8. For the people of the ancient world, the bottom line of it all was the ontology of chaos/cosmos, *physis/nomos*, insofar as development implied that a substance reached its plenitude. The current democratization is an ever-incomplete dynamic process which permanently runs the risk of inversion, of "de-democratization" (Tilly, 2010, p. 29). Otherwise, the underlying premise is now the idea of progress, a "synthesis of the past and a prophecy of the future," which means an unlimited expansion of the rational domain, more production, more economic growth, and more domination over nature. Nonetheless, this concept, which was at its height around 1900, had already started its decline before World War I, because faith in it had been lost (Nisbet, 1991, pp. 19–243). In any case, even though progress is not inevitable, this does prevent it from being possible (Wallerstein, 1997, p. 69 ff.).

The Decline of Democracy

Representative democracy, rejuvenated during the first modernity, also reached a stage of decadence or aging during the second one. Thus, the actual word "democracy," the features and the four perspectives which define it—constitutional, substantive, procedural, and process-oriented (Tilly, 2010, p. 37)—currently run the risk of being deprived of sense and becoming absolutely meaningless (Flores D'Arcais, 2013, p. 13). This is due to the fact that the three or four stages elapsed in the development of democracy—Liberal Rule of Law; Liberal-Democratic Rule of Law; Liberal-Social Rule of Law; and the post-material State—which represented a great historical pact between the forces of the market, those of labor and the democratic ones, find themselves in a situation of profound decline or crisis. The reason can be found, on the one hand, in three acute problems: ungovernability; the privatization of the public sphere; and the existence of an invisible power (Bobbio et al., 1985, pp. 12–14; Bobbio, 2007, p. 36 ff.; Requejo Coll, 1990, pp. 76–167); and, on the other hand, in the presence of external and internal enemies.

Thus, democracy is witnessing how the external risks or enemies (totalitarian regimes, fundamentalist movements) have been brought together with the internal and currently dominant ones (populism, ultraliberalism, and messianism) (Todorov, 2012, p. 8 ff.).

It can consequently be observed that, broadly speaking, these phases of contemporary democracy have experienced—like the Nation-State, the Western world's most important political creation (Barzun, 2001, p. 1142)—a passage from stability to crisis, and from vigor to decadence. Although the idea of "democracy as equilibrium" had prevailed until the 1960s (Macpherson, 1981) and it was widely accepted that this form of government brought a stable order and widespread prosperity with it, the faith in its potential was lost little by little, and democracy started being seen as incapable of finding justice- and equality-based solutions to the numerous problems and challenges that it is faced with (Roche Cárcel, 2013, p. 58). All this has been accompanied by the polarization of democratic ideals, since we currently witness a confrontation between the postulates according to which the State has gone into a crisis situation because it is overloaded with the opinion of those who believe that it has lost legitimacy after the break of class relations due to the limitations that Capital has imposed on Politics (Held, 1993, pp. 273–277).

The Generational Coexistence of Young and Old People in the 15-M Movement

The Spanish *15-M* movement is closely linked to the decay of representative democracy as well as to the overall crisis of contemporary society. May 15, 2011, witnessed the outbreak of a peaceful revolution promoted by Spanish youth, by 23-to-30-year-olds who felt frustrated because of the limited expectations that they had of becoming "middle-class workers." They consequently saw that their future was worse than that of the previous generations and additionally thought that those responsible for the crisis—the elites that punished them with permanent precariousness—were rescued from it thanks to public money. It all resulted, during the days immediately after May 15, in thousands of young people filling and occupying the squares of cities—the most important ones, Madrid and Barcelona—while social media transmitted their discontent. Therefore, both the squares and the Internet assumed a starring role as true agoras that staged political and social debates (Pastor Verdú, 2011, pp. 109–124; Fernández-Planells et al., 2013, p. 128 ff.).

Nevertheless, although 15-M can be defined as a youth movement, it is transgenerational too. As an example thereof, very broad as well as heterogeneous sectors adhered to the two essential components of this movement—"Real Democracy Now" and "Youth Without Future" (Errejón, 2011, pp. 121–122)—as a result of which the profile of participants became ideologically plural and diverse, apart from having an interclass nature (Sanz & Mateos, 2011, pp. 521–536), and, in short, its constitution was intergenerational (Pastor Verdú, 2011, p. 109 ff.). In fact, though the predominant profile was initially that of a young person, with the passing of time, the maturation of the movement, combined with the widespread dissatisfaction caused by the crisis that affects the whole of Spain, has allowed the youth to connect with increasingly large social sectors. It is not by chance that the challenge of 15-M—as stressed by its social actors themselves—precisely consisted in involving 99% of the population. Consequently, other generations have gradually joined the feeling of indignation, including the so-called "*yayoflautas*," who describe themselves as a group of people older than 50 belonging to the generation that fought to achieve a better life for their children and who now see that the future—both theirs and that of their descendants—is endangered. Therefore, three generations of grandparents, parents, and young children—four, if we add the children on strollers pushed by their parents—have joined the *Movimiento de Indignados* [Movement of Indignant People] under different slogans and with two shared objectives: fighting against social cutbacks; and changing the rules of the game (Pastor Verdú, 2011, p. 109 ff.; Fernández-Planells et al., 2013, p. 128 ff.).

The change demanded by the participants in this movement essentially materializes in their fight against "the system"—politicians, bankers, institutions, and mass media (Vallespín, 2011, p. 10)—and against the limitations of representative democracy (López Aguilar, 2011, p. 25), as clearly illustrated by their statements: "We are here because we want a new society which prioritizes life over economic and political interests. We advocate a change in society and social awareness."[1]

This profound transformation of life demanded by the 15-M Movement is consistently implemented by its adherents through their ways to protest and organize themselves, which means a "radical novelty," both in the practices of innovation in political participation (Errejón, 2011, p. 122) and in the collective actions undertaken. More specifically, they focus on the mobilization formats represented by camps and on updating their message, insofar as these demonstrators propose a radical break with the most usual organization methods—those of parties and unions, among others—because of their decentralized, transparent, self-managed, horizontal, and self-representative nature. These new formats additionally stand out for their acceptance and the broad social connections that they generate (Sanz & Mateos, 2011, pp. 532–536).

Summing up, the preservation—maintaining the rights achieved by the older ones—and renovation of democracy go hand in hand in this movement, and so does the coexistence of generations; it thus shapes a hybrid combination of youth and old age.

15-M is consequently another illustrative example of what is happening in contemporary society. This movement, which simultaneously includes two generations, accordingly shapes a mixture of youth and old age, which would explain why the creativity exhibited and characterized by the novelty and innovation of its proposals meant at the same time a return to the origin or, expressed differently, it simultaneously tried to keep the past and build the future with new elements, that is, to maintain the social achievements and to update democracy. However, it becomes visible in this movement that youth has lost a certain degree of power, that young people cannot renovate society on their own. No wonder, seeking to achieve their aims, they have had to join other population segments, especially the "*yayoflautas*" mentioned earlier. And this has affected the strength and depth of reforms. Therefore, it is no longer conflict or collision that characterizes the clash between generations described by

[1] See *Manifiesto de Acampadasol*: https://15mpedia.org/wiki/Manifiesto_de_Acampadasol. Consulted on January 1, 2020.

Sociology but the coexistence between them and the solidarity of young people with the old ones, and ultimately their mutual dependence.

3 Limits and Possibilities of Creativity to Overcome the Crisis of Representative Democracy

This last chapter has firstly served to define the terms "crisis" and "representative democracy," which allowed me to show the most important features in both of them. The aim was to know the role that could be played by creativity in providing an exit from this situation of crisis, particularly of representative democracy. Nevertheless, since creativity has an ambivalent and polarizing nature and keeps its conflicting options open—because it can be one thing and the other—this enables creativity either to perform an important dialectical and mediating function which encourages the connection or fusion of the different opposing poles or to separate them even to a greater extent, stressing their antagonisms, their contradictions, and their conflicts. Thus, if creative economy—see the previous chapter—has widened the divide between the market and the State, between politics and economy, between capitalism and representative democracy, and has even activated the contradictions inherent to capitalism, creativity has delved deeper into the crisis, the aging, the decay, and the collapse of democracy. Likewise, it has created new inequalities between rich and poor people, as well as between creative and noncreative States. In this context, creativity cannot be seen today only as a salvation before the multiple difficulties and uncertainties which do so much harm to representative democracies, since it is also—and we should not forget it—a source of intensification of those same problems and even generates other new and unforeseen ones. Hence, in sum, the need to address creativity in a critical way, discerning its risks, fragilities, and limits.

In addition to this, the interrelation of these three notions of crisis, representative democracy, and creativity evidences the polarization of democratic ideals, of creativity itself, and of the economic globalization logic. Furthermore, creativity—exactly like crisis and capitalism—highlights the contradictions, antagonisms, and conflicts, without forgetting that the crisis has resulted in the conceptual emptying (as it has also happened in the case of representative democracy) and has delegitimized—like democracy—the system of values, affecting social structure and change, in the same way as creativity has.

To this must be added that, because the global economic system has commoditized creativity, depriving it of meaning, and since creativity has been empowered as opposed to culture, it has thus contributed to aggravate the crisis of representative democracies and of their values. In the face of this situation, at least the following proposals can be made:

As has been repeatedly underlined from a variety of social contexts, the failure of politics makes it necessary to repoliticize education, economy, culture, and democracy, which could be done creatively, that is, taking creativity back to the sociopolitical sphere instead of focusing it exclusively on the economic environment.

The tension existing between the global scope of financial markets and the domestic sphere of Nation-States has confronted Capitalism with Representative Democracies. Within this tension, creativity has not solved the problem, as it has generated new inequalities, risks, relocations, invisibilities, and tax evasion tricks, which means that it has actually increased the distance between economy, culture, and democracy, and the one which separates the latter from capitalism. The need would consequently exist to reinvent the tension between democracy and capitalism (De Souza, n.d.) and, to that end, it would be advisable to reduce the separation existing between citizens and experts, between the State and Civil Society, between the global and the local, between representation and participation policies, or, expressed differently, we should creatively complement Representative Democracy with Participatory Democracy.

An emptying of meaning has taken place during the crisis that has especially hit representative democracies, which has led to confusion between creativity and contingency, to the vanishing of the borders between information, entertainment, and fiction in representative democracies, as well as to the replacement of interrelations by transactions in digital global economies. Therefore, leisure and work, reality and virtuality, passiveness and activity, creativity and consumption have become entangled as well. This ultimately explains why democracy—lacking in solid political foundations—is removed by global economy; and why culture, without any solid values, has been displaced by a commoditized, vanishing, dematerialized, virtualized, and transubstantiated creativity which has also been stripped of its substance. This means that the governance of creative cities and the correction of economistic and instrumentalist excesses arise as some of the most important challenges for cultural policies in the coming years (Sánchez Belando et al., 2012, p. 47). Therefore, the meaning of culture, creativity, and democracy should be resignified as connected—albeit different—terms, and creativity ought to be oriented to reinforcing the meaning of culture, while the latter should strengthen that of democracy.

The crisis has encouraged creativity to focus essentially on social change but, by prioritizing the most profitable novelties, it has commoditized change and fused it with the pace of goods. Consequently, creativity does not seem to have really helped solve the problems associated with the delegitimization of democracy and its values; it has apparently not pursued a reformist or revolutionary social transformation but the increase of economic profits for the leading industrial sectors, making them more powerful. In this way, it has become just another merchandise, as well as a polarizing and unequalizing instrument of the global capitalist system. Furthermore, we must remember that creativity not only transforms the systemic defects, giving rise to a new order, but also reinforces what survives of the existing one. This is why claiming the importance of creativity for Sociology means rethinking it in its connections with structure and change alike. And, if we do so, we will undoubtedly carry out an exercise of—sociological—imagination and creativity.

Bibliography

Adorno, T. W. (1991). *The culture industry*. Routledge.

Amin, S. (2002). *El capitalismo en la Era de la Globalización*. Paidós.

Arango, J. (2003). *Inmigración y diversidad humana. Una nueva era en las migraciones internacionales* (p. 268). Revista de Occidente.

Arendt, H. (1988). *Sobre la revolución*. Alianza.

Arendt, H. (1999). *Crisis de la República*. Taurus.

Ariès, P., & Armiño, M. (1987). *El hombre ante la muerte*. Taurus.

Attali, J. (1979). El orden por el ruido. El concepto de crisis en teoría económica. *Communications*, 25.

Baños, J. (2006). Teorías de la democracia: debates actuales. *Andamios, 2*(4, June), 38–39.

Barzun, J. (2001). *Del amanecer a la decadencia. Quinientos años de vida cultural en Occidente (de 1500 a nuestros días)*. Taurus.

Bauman, Z. (2003). *Modernidad líquida*. FCE.

Bauman, Z., & Bordoni, C. (2016). *Estado de crisis*. Paidós.

Beck, U., & Beck-Gernsheim, E. (2016). *La individualización. El individualism institucionalizado y sus consecuencias sociales y políticas*. Paidós.

Beetham, D. (1985). *Max Weber and the theory of modern politics*. Polity Press.

Beinstein, J. (2005). *El concepto de crisis a comienzos del siglo XXI*. Archivo Chile, CEME (Centro de Estudios Miguel Enríquez).

Béjin, A. (1979). Crisis de valores, crisis de medidas. In R. Starn et al. (Eds.), *El Concepto de Crisis*. ediciones Megápolis.

Bell, D. (1992). *Las contradicciones culturales del capitalismo*. Alianza.

Beltrán Llavador, J. (2014). Por una democracia creativa en educación superior: una tarea inacabada. *Revista del IICE, 36*, 5–19.

Bergua, J. A., Pac, D., Báez, J. M., & Serrano, C. (2016). La clase creativa. Una aproximación a la realidad española. *Revista Internacional de Sociología RIS, 74*(2, April–June), 1–19.

Bobbio, N. (2007). *El futuro de la democracia*. FCE.

Bobbio, N., Pontara, F., & Veca, S. (1985). *Crisis de la democracia*. Ariel.

Castoriadis, C. (1998). La polis griega y la creación de la democracia. In *Los dominios del hombre. Las encrucijadas del laberinto* (pp. 97–132). GEDISA.

Castoriadis, C. (2006). *Lo que hace a Grecia. 1. De Homero a Heráclito. Seminarios 1982–1983. La creación humana II*. FCE.

Castoriadis, C. (2008). *El mundo fragmentado*. Caronte.

Cristiano, J. L. (2010). La "Creatividad" de la Acción: la Teoría Joasiana y la cuestión de lo Imaginario. *Nómadas, Revista Crítica de Ciencias Sociales y Jurídicas, 25*, 1–19.

Cristiano, J. L. (2012). La creatividad como aspecto de un replanteamiento de la Teoría Sociológica de la Acción. *Cuadernos de Filosofía Latinoamericana, 33*(106), 33–62.

De Souza, B. (n.d.). Globalización y Democracia. *Ponencia presentada en el Foro Social Mundial Temático*.

Elias, N. (2012). *La Soledad de los moribundos*. FCE.
Errejón, I. (2011). El 15-M como discurso contrahegemónico. *Encrucijadas. Revista Crítica de Ciencias Sociales, 2*, 120–145.
Farfán, H. R. (1999). Ni acción ni sistema: el tercer modelo de acción de Hans Joas. *Sociológica*, año 4, número 40, *Perspectivas contemporáneas en la Teoría Social*, mayo–agosto, 35–63.
Fernández-Armesto, F. (1999). *Historia de la verdad y una guía para perplejos*. Herder.
Fernández-Planells, A., Feixa Pampols, C., & Figueroas-Maz, M. (2013). 15-M En España: Diferencias y Similitudes en las Prácticas Comunicativas con los Movimientos Previos. *Última década, Proyecto Juventudes, 21*(39), 115–138.
Flores D'Arcais, P. (2013). *¡Democracia! Libertad privada y libertad rebelde*. Galaxia Gutenberg-Círculo de Lectores.
Florida, R. (2010). *La Clase Creativa. La transformación de la cultura del trabajo y el ocio en el siglo XXI*. Paidós.
Freund, J. (1979). Observaciones sobre dos categorías de la dinámica polemógena. In R. Starn et al. (Eds.), *El Concepto de Crisis*. ediciones Megápolis.
Fumagalli, A. (2010). *Bioeconomía y capitalismo cognitivo. Hacia un nuevo paradigma de acumulación*. Traficantes de Sueños.
Habermas, J. (1984). *Ciencia y Técnica como ideología*. Tecnos.
Habermas, J. (1989). *El discurso filosófico de la modernidad*. Taurus.
Held, D. (1993). *Modelos de democracia*. Alianza.
Ingham, G. (2010). *Capitalismo*. Alianza.
Joas, H. (2002). *Creatividad, Acción y Valores*. Universidad Autónoma Metropolitana.
Joas, H. (2013). *La creatividad de la acción*. CIS.
López Aguilar, J. F. (2011). El hartazgo de las instituciones. *Claves de razón práctica, 215*, 20–28.
Macpherson, C. B. (1981). *La Democracia Liberal y su época*. Alianza.
Marangudakis, M. (2016). Visions of brotherhood; a comparative analysis of direct democracy in ancient and modern Greece. *Política y Sociedad, 53*(3), 773.
Mills, C. W. (1999). *La imaginación sociológica*. FCE.
Montesquieu. (1997). *Cartas persas*. Cátedra.
Morin, E. (1979). Para una crisología. In R. Starn et al. (Eds.), *El Concepto de Crisis*. ediciones Megápolis.
Morin, E. (1981). *El método I. La naturaleza de la naturaleza*. Cátedra.
Morin, E. (1994). *El hombre y la muerte*. Kairós.

Nisbet, R. (1991). *Historia de La idea del progreso*. Editorial Gedisa, SA.
Nussbaum, M. C. (1995). *La fragilidad del bien. Fortuna y ética en la tragedia y la filosofía griega*. Visor.
Pastor Verdú, J. (2011). El Movimiento 15 M. Un nuevo actor sociopolítico. *Sociedad y Utopía. Revista de Ciencias Sociales, 38*, 109–124.
Pope, R. (2005). *Creativity. Theory, history, practice*. Routledge.
Requejo Coll, F. (1990). *Las democracias. Democracia antigua, democracia liberal y Estado de Bienestar*. Ariel.
Roche Cárcel, J. A. (2013). La frágil construcción de la democracia en la Grecia Antigua. In the monograph *Transiciones. La fragilidad de la democracia*, issue coordinated by J. A. Roche Cárcel, *Res Publica. Revista de filosofía política*, No. 30, Year 16, 15–58.
Rodríguez Adrados, F. (1983). *La Democracia ateniense*. Alianza.
Sánchez Belando, M. V., Rius Ulldemolins, J., & Zarlenga, M. I. (2012). ¿Ciudad creativa y ciudad sostenible: un análisis crítico del "modelo Barcelona" de políticas culturales. *Revista Crítica de Ciências Sociais, 99*, 31–50.
Sánchez Capdequí, C. (2014). *El dinamismo de los valores. Crisis y creatividad en la sociedad moderna*. Anthropos.
Sanz, J., & Mateos, Ó. (2011). 15-M. Apuntes para el análisis de un movimiento en construcción. *Revista de Fomento Social, 66*, 517–544.
Serrano, C. (2014). *Sociología de la economía creativa en Aragón. Números, trayectorias vitales y entornos*. Doctoral Thesis, Universidad de Zaragoza, Zaragoza.
Soros, G. (1999). *Crisis del Capitalismo Global*. La. Sudamericana.
Starn, R. (1979). Metamorfosis de una noción. Los historiadores y la "crisis". In R. Starn et al. (Eds.), *El Concepto de Crisis*. ediciones Megápolis.
Thom, R. (1979). Crisis y catástrofe. In R. Starn et al. (Eds.), *El Concepto de Crisis*. ediciones Megápolis.
Tilly, C. (2010). *Democracia*. Akal.
Todorov, T. (2012). *Los enemigos íntimos de la democracia*. Galaxia Gutenberg-Círculo de Lectores.
Valencia López, H. (2016). La política y la democracia como creaciones imaginarias: de los griegos a nosotros. *Atenea* 513, I Sem., 125–135.
Vallespín, F. (2011). La fatiga democrática. *Claves de razón práctica, 215*, 10–18.
Vattimo, G. (1995). *El fin de la modernidad. Nihilismo y hermenéutica en la cultura posmoderna*. GEDISA.
Vernant, J.-P. (2003). *El universo, los dioses, los hombres. El relato de los mitos griegos*. Anagrama.
Wallerstein, I. (1997). Incertidumbre y creatividad. *Espacios*, pp. 69–71.

Part IV

Final Coda: Broken Creativity

In conclusion, the temporal narrative that constructs creative originality has been dislocated in late modernity and, with it, creativity has been broken. First, because currently there is no space or time; second, because it has gone from seeking fusion, characteristic of the categories of origin, to fragmentation and inequality, specific of the originary categories; and, finally, because creativity has ceased to seek the survival of society and to want to escape existentially from time—as happened in ancient myths—and has ended up consuming time and being consumed by it, in the originary categories of aging. This process of transformation of creativity has followed two phases in modernity: child and youth rejuvenation of the first and senilism of the second. The latter has been caused, basically, by the role of the world wars, the demographic aging of society, the senescence of capitalism, contemporary art and science, and the crisis of representative democracy.

This generalized creative senility raises many doubts about the possibility that creativity, as suggested by Hans Joas, could represent, at least in the short term, as a way out of the contemporary crisis. On the contrary, it becomes a symptom of that crisis, rather than an opportunity for structural transformation of society. This book insists that the rupture of the temporal narrative that has traditionally been caused by creative originality and that connected the past, the present, and the future could be at the base of that profound transformative impossibility that creativity embraces today. Perhaps, only by returning to History, that is, by re-establishing creativity solidly established in the past, critically revisited, and at the time preserved and transformed, is it possible to experience with intensity and renewed energy a present that, in itself, becomes plural. Finally, from it, the possible trends of the future could be drawn.

9

Creativity Without Time

1 The Path Toward Creativity with Neither Space Nor Time

Phases of Civilization

Numerous proposals about mankind's development have been made in the field of Social Sciences. Broadly speaking, they can be divided into two main strands: the one which prioritizes the evolution toward the individual sphere; and, on the contrary, that which stresses the community- or society-related aspects within this process. Among those belonging to the former stands out the one described by M. Weber—seen in Chap. 2—since the German sociologist argues that the three phases through which mankind develops finish with the triumph of individualization: the tribal societies in which an ethics of neighborhood prevails, the prophetic or salvation era dominated by the ethics of fraternity, and the individualistic one of the modern capitalist or rationalist society. In the latter, there is consequently no longer an ethics of neighborhood or fraternity but an egoistic interest, an individualistic competition, and a

J. A. Roche Cárcel, *Creativity and Time: A Sociological Exploration*, Palgrave Studies in Creativity and Culture, https://doi.org/10.1007/978-3-030-84838-5_9

mechanical, commoditized, cold, and purpose-oriented rationalization of life which locks individuals—devoid of emotions—inside an iron cage.

Following along the same lines, the sociologist Shmuel Eisenstadt (2017, p. 85) situates reason at the top of social evolution, since he considers that Modernity is the result of three impulses: the naturalization of the human being, society, and nature; the publication of autonomy; and the potential supremacy of reason in the construction of the world. Well, it is the last of these stages that places the emphasis on human autonomy, on that of reason and/or will. The individual—as Schopenhauer (2017) would say—therefore believes that the world is "my" will and "my" representation.

Likewise, referring to the historical development of Greek philosophy, Robert Bellah (2017, p. 475) explains that such development goes through three phases: the first one represented by Parmenides and Heraclitus (the creators of the objective being); the second one embodied by Protagoras and the Sophists, who make a turn toward subjectivity; and the third one characterized by Socrates and Plato, who return to the subjective reason by incorporating the objective moment.

In a similar vein, the Spanish philosopher José Ortega y Gasset believes that, of the three "attitudes" toward being, the last one would precisely have an individual nature. Indeed, the first one, that of the Greek world and the medieval universe, shows a realistic vision according to which being lies outside humans. The second one, associated with Modernity, is characterized by idealism, that is, by the fact that being finds itself inside one's thoughts. This stage begins in the seventeenth century, the time of Descartes, when the modern human being reaches his plenitude and wishes to break with what is old, since the individual feels that he has come of age and is fully responsible for his thoughts and the moment when he thinks that he can find himself. Finally, the third one defines being as the essence, but the latter is absent now, it is not within sight; hence the need to look for it and to do it (Ortega y Gasset, 2008, pp. 219 and 233–5). Consequently, it follows from Ortega's views that this pursuit and construction must fundamentally take place in the human interior rather than outside the human being, in the world or in society, and, therefore, in whatever might define the individual as such.

There are also conceptions which lay the emphasis on the societal sphere as opposed to the individual one. An example of them is provided by Arnold van Gennep, who thinks that rites of passage are built around three successive phases: separation; marginality; and aggregation (Bauman & Leoncini, 2018, p. 44). Above all, these consecutive stages highlight a center—society—from which the individual emerges. This individual firstly distinguishes himself and moves away from society, and ends up cornered, and is finally incorporated again after complying with the collectively determined rules and rites. Consequently, prior to returning to the group, the individual has socialized and interiorized his society's norms, habits, rituals, traditions, symbols, and procedures, making them their own; only then will he be ready to go back into that social context. However, when he does so, he is no longer that transitory individual, since he has become part of society.

The victory of what is shared can likewise be identified in the sociologist K. Mannheim, according to whom there are three main historical stages in mankind: horde solidarity; individual competition; and transindividual group solidarity (Mannheim, 2018, p. 77). And once again, society appears as the axis, even though the last phase recognizes the existence of individuals who, despite maintaining that condition, have left competition behind and found what is common.

Even though these different evolutions of civilization can largely help me develop the phases which, in my view, sequence the development of creativity, I have left for the end the civilizing development proposed by Merlin Donald, as it is the one which comes the closest to the time narrative of creativity that underpins this book.

According to Donald (2017, pp. 50–54), human culture is deployed in four successive strata of primate/hominid culture evolution, determined by the brain–culture coevolution: episodic; mimetic; mythical; and theoretical. Strictly speaking, the great cultures are the last three. A prelinguistic utilization exists in the mimetic one, while the mythical one resorts to metaphors which explain and regulate every single aspect of life, which is permeated by myth and narrative, one of the three basic modes of thinking, the way in which we understand existence, personal and collective identities, as well as the source of ethics, politics, and religion. Archaic religions represent highly developed cultural advances with respect to social tribes, inseparably linked to the political organization of the primitive State. More specifically, both the mimetic culture and the mythical one probably mutated in the richest areas of the house and the harvest, while population density simultaneously generated cultural innovation. Otherwise, the mythical culture is older than the theoretical one and continues to be essential for our relationship with the world at present. The theoretical culture is finally added to the mythical and the mimetic ones, insofar as it establishes a dialog with both of them with the aim of modeling the human universe in its entirety. In other words, it represents a significant process which is preserved, and which has consequently never been lost, although it has indeed been reorganized and triggers new substantial processes that promote an adaptive or innovative aspect variation. Furthermore, its key elements were gradually implemented, namely, graphic invention; external memory; and theory construction. In sum, its outcome has been "the cultural hybrid system," a new compound synthesis of the mimetic, the mythical, and the theoretical (Bellah, 2017, pp. 109–488).

The process described by Donald is of special interest for this essay, insofar as the theoretical culture becomes a critique and an alternative to the prevailing ones, insofar as it creates new reality models while simultaneously maintaining the old ones. That is to say, the nascent state of creativity, the ultimate result of a long bio-cultural evolution, is supported and inserted in a historical framework by mimetic and mythical aspects. Hence, this development fits in very well with the time narrative of creativity depicted in Chap. 1.

On the other hand, the historical stage to which Donald refers would have been reached in the time-axis described by the German psychiatrist and philosopher K. Jaspers, though the last phase that he mentions could suggest a questioning of the relevant role played by creativity in the development of a spiritual and not merely technological society. Thus, according to Jaspers, there are four occasions on which human beings have originated a new base for their existence: Prehistory; the foundation of the oldest major cultures; the "time-axis," an "open state"—to which I dedicated Chap. 4; and the technical–scientific age. Nevertheless, the last one, ours, has a nature of technical transformation rather than one of "eternal spiritual creation" (Jaspers, 2017, p. 49 and 209).

Consequently, creativity would have become technified and abandoned its temporal dimension, the narrative rooted in the origin that is in turn the seed of the originating; therefore, this last dimension, the future, ends up losing its substance and legitimacy, and becomes weakened and diluted.

Otherwise, the French sociologist M. Duvignaud—agreeing with the ideas of Jaspers and Weber in this regard—also concludes that the present has stopped being creative in the way it used to be. After all, in his view, ever since the eighteenth century, Europe has turned its back on the squandering of Baroque forms, gold and color; and has gradually moved away from creative play by committing itself to profitability and efficacy. We have therefore gone from a play civilization to one oriented toward profitable production, although it recognizes that human experiences continue to exist which are not exhausted in productive rationality (Duvignaud, 1980, pp. 86–91).

Stages of Creativity

As I see it, the first stage of creativity would begin at the very moment that the human being arises as such at an undefined date—if that date can be ever specified—related to, and at the same time undifferentiated from, the animal world. In the course of that wonderful inaugural event, the origin of human creativity, the cultural world generated by the latter adopts a certain autonomy–dependence with respect to biological evolution. Furthermore, this initial phase, which would last until the creation of the earliest States after the Neolithic revolution, would be the one of objectivity, characterized by the presence of a strange world which amazes and scares the human being and is the source of his creativity, fundamentally stimulated by the ability to survive when faced with that fierce world.

A second stage could be represented by the axial civilizations of the first millennium defined by Jaspers which still shape our time, when we add to the awareness of being part of a world designed by the ancestors the possibility to question the inheritance received and even to willfully originate a society other than the previous one and a self-conscious individual who is also aware of his possibilities and his duties, and is accordingly detached and free from the ancient collective ties. A third stage was going to arise in the seventeenth century, the time when this individual became truly empowered and acquired consciousness of the fact that his own will—and the creativity that the latter formalizes—can build and dominate nature and the world.

Finally, the last period, the modern one strictly speaking, would consist of two substages. During the first modernity, individual autonomy will be placed at the service of rationality, subject to purposes of the capitalist economic system, and of his first-born child, the scientific technical system. Nonetheless, the true individual autonomy demanded by the enlightened project and which was going to allow the human being to come of age is constrained by that capitalist–technical–scientific system which hinders the capacity to rid the human being from the old and new shackles by means of creativity. Instead, an individual has emerged who is as relative as the space and time he lives in, because he can no longer be sure either about his role of transforming agent or about his function as the observer of a world and a society which are vanishing in front of his eyes. Thus, the individuals who form part of society—with a deep desolation, with the feeling that they are locked in a new prison and without any chances to escape from it, and, in sum, with the conviction of having been exhausted and impoverished by the successive wars—see how their expressive potential and creative potential no longer manage to play a prominent role in the change of their society and their civilization, both of which fall into decline. This state becomes enhanced and institutionalized in the second modernity, when the dematerialization of economy and, with it, that of space and time, situates individuals in non-places, in virtual spaces, and in instantaneous, ephemeral, and accelerated times and events. Furthermore, this second modernity has witnessed the establishment of individualization as a destination formed by individuals left (Heidegger, 2003, p. 197) to their own devices in a world without limits, an undetermined and chaotic one through which they float as if they were angels, albeit without their omnipowers, to which is added that they have to assume all the risks, both theirs and those of the institutions.

Consequently, after having lost the conviction that it is possible to build the future that will serve to overcome this critical stage, creativity has now come to be determined by crisis, contingency, risk, and uncertainty.

To sum up, we could conclude that Western civilization has stopped observing the world objectively (Bellah), trying to imitate it as effectively as possible, and has begun to do so in a subjective way, firstly with a strong rational will and, secondly, with a reflexive rationality that constitutes an attempt to create the new, starting from the human psyche, from that location where complex interferences exist between the inside and the outside, between reason and emotions. However, this journey leaves the outer world to penetrate the inner human mind, and it does so at the same time that creativity changes its space: the world is no longer the object of creativity now; instead, the emphasis is laid on the creation process. An artist's conception is certainly an internal process, so much so that, when he creates, he is in ecstasy, "outside himself," so that he can only create his imaginary world by forgetting about the real world (Zweig, 2015, pp. 17–21). And the result is that the modern artist becomes "too aware" of his work (Gadamer, 1996, p. 13).

The time of creativity simultaneously changes, since creativity no longer clearly distinguishes the past, the present, and the future or tries to intertwine them in a time narrative. It has broken its bridges with space and time and is now forced to focus on contingency, risk, and uncertainty—a sensation that appears to be more mental than real.

In short, creativity without time or space does not seem the most suitable one, neither to intensely experience a dislocated and excessively accelerated present nor to successfully design the promising future of our societies. Firstly because, if the creator is currently more interested in the creative process than in the objective world, the latter, its problems and the solutions to them, stop being a priority, and the willingness to transform becomes weakened. Secondly, because, once creativity has basically become a subjective mental process, its success will depend on whether it can be implemented in practice, on making it known to society, and on whether the latter accepts it or not. However, it must of course do so within the unfavorable environment of an individualization society where each individual seemingly pursues their own interest, without having or finding the emotional and rational solidarity required to integrate them into the collectivity. And thirdly, because a broken creativity can only give rise to a fractured, decomposing world whose parts find it difficult to stay holistically united.

Hence, our insistence on the need to reconsider the possibility of returning to a time narrative of creativity which can once again interrelate the past, the present, and the future, and which can create the originary from the origin. Only in that way—we think—will creativity be able to recompose its own broken fragments and those of the society which keeps them and perform, with guarantees of success, a mediating and fusing function between human beings and their creativity and the world.

2 From Fusion to Fragmentation

Creativity and *Coincidentia Oppositorum*—In the Imaginary

As a matter of fact, creativity is a single whole which cannot be divided into different fields of specialization. For instance, if science generated a highly significant creation—something that happened in the first modernity, as we explained earlier—it would be inevitably accompanied by other creative waves arising in various cultural and social contexts and, broadly speaking, every walk of life would be affected (Bohm & Peat, 2007, p. 214). It comes as no surprise then that the creator is a specialist who, thanks to his work, can unify his hybrid or mixed provenance, even if in a precarious way (López Mondéjar, 2015, p. 32).

The time narrative of creativity, which characterizes it, and which has been described here, also implies an effort to fuse; not in vain, time fulfills coordinating and integrating sociological functions. After all, it acts as a symbol of a huge network of relationships where the processes of different levels—individual, social, and nonhuman natural—are interwoven; we must remember that social and physical time, that of society and the one of nature, are inseparable. Therefore, determining time represents a merging activity, the implementation of a synthesis (Elias, 2013, pp. 39, 74–77 and 108), to which must be added that the unity of History is a goal rather than a factual state (Jaspers, 2017, p. 375).

It follows from this that culture, the great typically human creation which develops over time, constitutes a huge research task (Ortega y Gasset, 2008, p. 145), as well as a goal of humanization, of becoming truly human, in the same way that the creation of time established by creativity ultimately implies the culmination of the creating process rather than a fact in itself. Therefore, all cultural activities, particularly the most creative ones, express this unification. By way of example, play (essential in culture) beats the dichotomies of the split human life—matter and form, feeling and intellect, reality and necessity (Bellah, 2017, p. 691). Likewise, the symbolism of art precisely rests upon an insoluble game of opposites which are shown and simultaneously concealed (Gadamer, 1996, p. 87).

This fusion also takes place in myths and the ritual—sons of play. More specifically, mythical thinking sees *coincidentia oppositorum* as its peculiar vital element (Cassirer, 1989, p. 54), whereas rite triggers a change of state where antinomies dissolve (Balandier, 2014, p. 29). Even, in some rites, dances are perfectly coordinated and reach a harmonious fraternity which reconstructs the whole society. Not surprisingly, the context where they occur is one of hierarchical indifference and of a transitory reproduction of the old equality (Bellah, 2017, p. 694).

Similarly, at the carnival, everything is sanctioned by the union of opposites, particularly of the sacred and the clownish, inversion arising as its main operator. Inversion is actually the one which transforms hierarchies into masks, the one which breaks temporary obligations, censorships, and conveniences, as well as that which metamorphoses scarcity into plenty and consumption into consummation (Balandier, 2014, p. 120).

This is achieved in language, in science, in art, and in music too, even if it is momentary, the spell of the return of the lost unity. When it comes to language and music, for instance, a characteristic high degree of coincidence exists, oriented in this last activity toward the union between the self and the world, as well as toward the building of a sense of space without distinguishing places and time where the past and the future coexist with the present (Bellah, 2017, pp. 183–198).

In scientific creativity, it is the concept of metaphor that helps us evidence its fusing nature; not in vain, metaphor converts different—even unfathomable—concepts into equivalent ones. For that reason, it is a creative form where the key is the equalization of two very different things which allows it to enter a highly perceptive state of great energy and passion. In this regard, it comes as no surprise that contemporary science (see Chap. 3) has transformed space and time into interchangeable dimensions in such a way that the latter becomes space and the former is transformed into time (Bohm & Peat, 2007, pp. 45–88).

Creativity accordingly shapes a space and time which are fused and interchangeable, additionally converting them into the imaginary home—experienced as comforting and pacifying—of human beings uprooted by nature and thrown into the time of death. Nonetheless, a powerful energy emanates from that fictitious universe—as indelible and fragile as human beings themselves—which, by restoring the lost original unity, serves to give rise to a settled and legitimate future, the true goal sought with the construction of human time that all creativity entails.

This is at least what has basically happened in the categories that I have referred to as those of the origin, the ones which return to the beginning—the myths of creation, play, order and chaos, and axial civilizations—since a time narrative has been established in them where the origin and the originary, the imaginary and reality, have walked together.

Creativity Establishes Hierarchies, Unequalizes, and Differentiates—In Social Reality

Even though that imaginary level where it seems possible to achieve social and cultural unity, and even the cosmic one, appears to remain valid in the late modern actuality, this objective is not so easy to achieve in its factual reality. In fact, as has been verified throughout this book, above all in the originating categories—capitalism, creative industries, creative cities, work, and representative democracy—which best express the desire for utopia, to generate a future, creativity intensifies the opposite: inequality, hierarchization, and differentiation.

This is what happens, by way of example, with work which, in addition to being generally despatialized and dematerialized, uproots the worker. This likewise applies to the creators who develop their professional activity in creative industries, the best paid ones, in the leading cities in terms of creativity which increases their gross domestic product far above the urban areas located in the periphery or that are not so "creative." The disintegration of the social forces which supported representative democracy pressed by the centripetal energies of their internal and external enemies provides another representative example for the break of unity in contemporary society. Finally, this overall disintegration has a more acute impact on the peripheries of the capitalist system, where poverty and the uneven distribution of income are not negative effects caused by specific circumstances or erroneous policies, but result from the actual logic of the capitalist system and of the global polarization that is immanent to it (Amin, 2002, p. 31).

To cut a long story short, we currently have in front of our eyes a world and creativity which are both broken—shattered into a plethora of scattered, unconnected, and disoriented pieces. Creativity itself has moved further away from society and from itself, since it has fragmented the old unity between the imaginary and reality, each of which follows its own path now.

3 Creativity and the Existential Escape of Time

Certainly, the *flâneur* provides a very faithful description of the social agent who is characteristic of broken modernity. Charles Baudelaire defines him as that passionate observer who has his home "in the number, in the undulating, in movement, in the fugitive and the infinite" and who is consequently away from home. However, he feels like a citizen of the world who is everywhere, where he obtains pleasure from his unknown position (Baudelaire, 2004, pp. 86–87), from his nomadism, and from his fugacity. It seems advisable to remember in this sense that the indistinct, chaos, and unlimited contingency precisely constitute the creative core (De Carolis, 2017, p. 58).

We should add to Baudelaire's contribution, though, that the pleasure experienced by the modern citizen before a chaotic world is also accompanied by the fear of one's own expiration, the fear of death, the personal reason that impels human beings—after any transformation or flow—to discover something eternal and long-lasting (Elias, 2013, p. 146). After all, "[a]lthough they have to die, human beings have not been born to do that, but to innovate" (Ricoeur, 2010, p. 637); hence the significant value that they assign to symbolic order, precisely because it performs the specific function of denying contingency (De Carolis, 2017, p. 58).

From the beginning of time, myth, religious beliefs, and aesthetics have been the fundamental instruments which have allowed human conscience to experience the liberation from biological and historical time, that is to say, from death. And they have done so by constructing a fiction which they regarded as true and which represented a pragmatic metaphor of eternity. Without those instruments, that is, without myth, beliefs, and the arts, the human psyche would probably find itself unprotected when faced with its personal extinction and might even fall into despair and madness (Steiner, 2002, p. 79 and 264).

Myth, philosophy, literature, and art in general—and music in particular—very effectively carry out this function of fighting against the time of expiration, especially because they need a temporal dimension where they can be displayed, though with the aim of denying it. This is, for example, what literature does; it denies time and narrates the self to represent how it faces the main problem of life as a useful way to move forward despite the suffering (Bellah, 2017, pp. 72–75). Likewise, any musical piece suppresses time, turning it into an independent phenomenon (Steiner, 2002, p. 80). In a similar vein, the anthropologist Levi Strauss argues that it is characteristic for music, myth, and play to occur in a time outside time, a sort of extinction thereof which precisely takes place within a relaxed atmosphere, that of play, with which we can return to a point prior to biological time, to a form of life which is not subject to the fight for existence (Bellah, 2017, p. 191 and 691). Neither should we forget that artistic play seeks to overcome time, insofar as—the same as holiday—it entails a transitory celebration period which brings together all participants and means a break of the present, an escape from linear time, and, in short, a pursuit of eternity.

For that reason, the essence of art consists in getting rid of time and returning to the origin—Heidegger (Gadamer, 1996, pp. 22–3 and 118)—a notion into which F. Schiller brilliantly delves deeper when he writes that "[o]nly the aesthetic state is a whole in itself, because it combines all the conditions related to its origin and its duration. Only in it do we feel as though we were outside time" (Schiller, 1990, p. 295).

Consequently, the main task that art must perform today adopts the form of a paradox, since the artist resorts to time, creates his time (Zweig, 2015, p. 35), though to deny it. This is also mentioned by F. Schiller in letter 27 of *Aesthetic Education* (1990, pp. 133–34) when he points out that "the sensitive impulse needs changes and time to have a content; the formal impulse requires the extinction of time and the absence of changes. Therefore, the impulse in which both combine—the play impulse—will point at the extinction of time in time and at the reconciliation of evolution with the absolute being, from change to identity."

Nevertheless, what this actually expresses—as I had anticipated a few lines earlier—is the very human desire to achieve immortality, a "manufacturing" that distinguishes it from death, which is a fact (Bauman, 2007a, p. 15). Indeed, any game is defined by the suppression of time in time (Gregori, 2002, p. 48), but artistic play goes beyond this impulse, since it tries to "capture the permanence of the fugitive" (Gadamer, 1996, p. 22). Furthermore, deep inside, what it tries to do is prevent or hinder mankind from forgetting itself, that is its nature, its own mortality, which explains why, for art, death is neither a technical problem nor an ordinary one, as the former conversely represents its *raison d'être*, its cause, and its purpose. In this respect, as Z. Bauman rightly asserts, "thanks to art, once and again death is reduced to its true dimension: it is the end of life but not the limit of human nature" (Bauman, 2007b, pp. 15–17). After all, creation, which is a "miracle," forces the limits where our life infallibly tends to be locked up and ultimately seeks to overcome human mortality (Zweig, 2015, p. 14)—unsuccessfully, of course.

With the time narrative of creativity, with its originality that refers us back to the origin and to the originary, on the one hand, human activity is limited; that narrative places it between two milestones: that of the beginning and that of the goal or the end. But also, on the other hand, life is opened toward the future, whose lack of a prefigured or predefined form makes it become a metaphor of the desire for infinity and immortality.

In short, creativity allows the flow of life to become form—while simultaneously informalizing it—and to expand it, not toward the path of death but beyond it. In my view, that is where the anthropological dignity of creativity lies and, if we are able to maintain that stimulating plan, this time which is so lacking in time, decadent, critical, and aged might not end in itself, but could instead be updated and continue toward an indefinite and hopeful future.

Bibliography

Amin, S. (2002). *El capitalismo en la Era de la Globalización*. Paidós.

Balandier, G. (2014). *El desorden. La teoría del caos y las ciencias sociales*. GEDISA.

Baudelaire, C. (2004). *El pintor de la vida moderna*. Colegio Oficial de Aparejadores y arquitectos Técnicos, Consejería de Educación y Cultura y Fundación CAJAMURCIA.

Bauman, Z. (2007a). *Tiempos líquidos*. Tusquets.

Bauman, Z. (2007b). *Arte líquido*. Sequitur.

Bauman, Z., & Leoncini, T. (2018). *Generación líquida. Transformaciones en la era 3.0*. Paidós.

Bellah, R. N. (2017). *La religión en la evolución humana. Del Paleolítico a la era axial*. CIS.

Bohm, D., & Peat, F. D. (2007). *Ciencia, orden y creatividad. Las raíces creativas de la ciencia y la vida*. Kairós.

Cassirer, E. (1989). *Esencia y efecto del concepto de símbolo*. FCE.

De Carolis, M. (2017). *La paradoja antropológica. Nichos, micromundos, disociación psíquica*. Editorial Quadrata/Iluminuras.

Donald, M. (2017). Una aproximación evolutiva a la cultura. Consecuencias del estudio de la época axial. In C. Sánchez Capdequí (Ed.), *La creatividad social: narrativas de un concepto actual* (pp. 47–73). CIS.

Duvignaud, J. (1980). *Le jeu du jeu*. Bolland, Paris. Mesa redonda. Nuevos horizontes. El juego. p. 118. Retrieved March 13, 2020 from www.libertario.arte.bo/biblioteca/Mesa%20redonda%20Nuevos%20Horizontes%20-%20Jean%20Duvignaud%2C%20el%20juego%20del%20juego.html

Eisenstadt, S. N. (2017). La dimensión civilizacional en el análisis sociológico. In C. Sánchez Capdequí (Ed.), *La creatividad social: narrativas de un concepto actual* (pp. 73–95). CIS.

Elias, N. (2013). *Sobre el tiempo*. FCE.

Gadamer, H.-G. (1996). *La actualidad de lo bello*. Paidós.

Gregori, E. (2002). Individuo y naturaleza en Schiller. La noción de juego. *Materia, 2, Naturaleses*, 35–55.

Heidegger, M. (2003). *Ser y tiempo*. Trotta.

Jaspers, K. (2017). *Origen y meta de la historia*. Acantilado.

López Mondéjar, L. (2015). *Una espina en la carne. Psicoanálisis y creatividad*. Psimática Clínica.

Mannheim, K. (2018). *El hombre y la sociedad en la época de crisis*. CIS.

Ortega y Gasset, J. (2008). *En torno a Galileo*. Revista de Occidente-Alianza.

Ricoeur, P. (2010). *La memoria, la historia, el olvido*. Trotta.

Schiller, F. (1990). *Kallias; Cartas sobre la educación estética del hombre*. Anthropos.

Schopenhauer, A. (2017). *El mundo como voluntad y representación*. Alianza.

Steiner, G. (2002). *Gramáticas de la creación*. Siruela.

Zweig, S. (2015). *El misterio de la creación artística*. Sequitur.

Bibliography

Adorno, T. W. (1991). *The culture industry*. Routledge.

Alexander, J. (2017). Trauma cultural e identidad colectiva. In C. Sánchez Capdequí (Ed.), *La creatividad social: narrativas de un concepto actual* (pp. 315–347). CIS.

Amin, S. (2002). *El capitalismo en la Era de la Globalización*. Paidós.

Amin, S. (2005). *Más allá del capitalismo senil. Por un siglo XXI no norteamericano*. Editorial Paidós.

Anuri, D. (2001, November). La Democracia reforzada por la creatividad. *NUNTIUM*, 119–124.

Arango, J. (2003). *Inmigración y diversidad humana. Una nueva era en las migraciones internacionales* (p. 268). Revista de Occidente.

Arendt, H. (1988). *Sobre la revolución*. Alianza.

Arendt, H. (1999). *Crisis de la República*. Taurus.

Arendt, H. (2005). *La condición humana*. Paidós.

Argan, G. C. (1998). *El arte moderno. Del Iluminismo a los movimientos contemporáneos*. Akal.

Ariès, P., & Armiño, M. (1987). *El hombre ante la muerte*. Taurus.

Ariño, A. (2016). *La secesión de los ricos*. Galaxia Gutenberg.

Aristóteles. (1994). *Metafísica*. Gredos. Traducción Tomás Calvo Martínez.

Aristóteles. (2000). *Política. Introducción general de Miguel Candel Sanmartín y traducción y notas de Manuela García Valdés*. Biblioteca Básica Gredos.

J. A. Roche Cárcel, *Creativity and Time: A Sociological Exploration*, Palgrave Studies in Creativity and Culture, https://doi.org/10.1007/978-3-030-84838-5

Arizpe, L. (2001). Cultura, creatividad y gobernabilidad. In *Estudios Latinoamericanos sobre Cultura y transformaciones sociales en tiempos de globalización* (pp. 31–48). CLACSO, Consejo Latinoamericano de Ciencias Sociales.

Attali, J. (1979). El orden por el ruido. El concepto de crisis en teoría económica. *Communications, 25*, 86–100.

Balandier, G. (2014). *El desorden. La teoría del caos y las ciencias sociales*. GEDISA.

Baños, J. (2006). Teorías de la democracia: debates actuales. *Andamios, 2*(4, June), 38–39.

Barzun, J. (2001). *Del amanecer a la decadencia. Quinientos años de vida cultural en Occidente (de 1500 a nuestros días)*. Taurus.

Baudelaire, C. (2004). *El pintor de la vida moderna*. Colegio Oficial de Aparejadores y arquitectos Técnicos, Consejería de Educación y Cultura y Fundación CAJAMURCIA.

Bauman, Z. (2003). *Modernidad líquida*. FCE.

Bauman, Z. (2004). *La sociedad sitiada*. FCE.

Bauman, Z. (2007a). *Tiempos líquidos*. Tusquets.

Bauman, Z. (2007b). *Arte líquido*. Sequitur.

Bauman, Z., & Bordoni, C. (2016). *Estado de crisis*. Paidós.

Bauman, Z., & Leoncini, T. (2018). *Generación líquida. Transformaciones en la era 3.0*. Paidós.

Baumer, F. L. (1985). *El pensamiento europeo moderno. Continuidad y cambio en las ideas, 1600–1959*. FCE.

Beck, U. (2006). *La sociedad del riesgo global*. Siglo XXI.

Beck, U. (2007). *Un nuevo mundo feliz. La precariedad del trabajo en la era de la globalización*. Paidós.

Beck, U., & Beck-Gernsheim, E. (2008). *El normal caos del amor. Las nuevas formas de la relación amorosa*. Paidós.

Beck, U., & Beck-Gernsheim, E. (2016). *La individualización. El individualism institucionalizado y sus consecuencias sociales y políticas*. Paidós.

Beetham, D. (1985). *Max Weber and the theory of modern politics*. Polity Press.

Beinstein, J. (2005). *El concepto de crisis a comienzos del siglo XXI*. Archivo Chile, CEME (Centro de Estudios Miguel Enríquez).

Béjin, A. (1979). Crisis de valores, crisis de medidas. In R. Starn et al. (Eds.), *El Concepto de Crisis*. ediciones Megápolis.

Bell, D. (1992). *Las contradicciones culturales del capitalismo*. Alianza.

Bellah, R. N. (2005). What is axial about the axial age? *Archives Européennes de Sociologie, XLVI*(1), 69–87.

Bellah, R. N. (2017). *La religión en la evolución humana. Del Paleolítico a la era axial.* CIS.

Beltrán Llavador, J. (2014). Por una democracia creativa en educación superior: una tarea inacabada. *Revista del IICE, 36*, 5–19.

Beltrán Villalva, M. (1985). Cinco vías de acceso a la Realidad Social. *REIS, 29*, 7–41.

Beltrán Villalva, M. (2012). Sobre hermenéutica: de la Filosofía a la sociología empírica. *RES. Revista Española de Sociología, 17*, 9–26.

Beltrán Villalva, M. (2016). *Dramaturgia y hermenéutica: para entender la realidad social.* CIS-Centro de Investigaciones Sociológicas.

Benéitez, B. (2005). La ciudadanía de la democracia ateniense. *Foro Interno, 5*, 37–58.

Bergua, J. A. (2015). *Postpolítica. Elogio del gentío.* Biblioteca Nueva.

Bergua Amores, J. Á., Serrano Martínez, C., Báez Melián, J. M., & Pac Salas, D. (2016). *La clase creativa. Una aproximación a la realidad española* (No. ART-2016-95206).

Bergua, J. A., Carretero, E., Báez, J. M., & Pac, D. (2016a). *Creatividad. Números e imaginarios.* CIS.

Bergua, J. A., Pac, D., Báez, J. M., & Serrano, C. (2016b). La clase creativa. Una aproximación a la realidad española. *Revista Internacional de Sociología RIS, 74*(2, April–June), 1–19.

Beriain, J. (1999). Del reino de Jaungoikoa (Dios) al politeísmo moderno. In J. Beriain & R. Fernández Ubieta (Eds.), *La cuestión vasca* (pp. 70–86). Anthropos.

Beriain, J. (2000). *La lucha de los dioses en la modernidad. Del monoteísmo religioso al politeísmo cultural.* Anthropos.

Beriain, J. (2008). *Aceleración y tiranía del presente. La metamorfosis en las estructuras temporales de la modernidad.* Anthropos.

Beriain Razquin, J. (2017). La metamorfosis del don: ofrenda, sacrificio, gracia, substituto técnico de Dios y vida regalada. *Política y Sociedad, 54*(3), 645–667.

Bilbeny, N. (2012). *Ética.* Ariel.

Billinghurst, J. (2007). *Mujeres tentadoras.* Océano.

Blanco Freijeiro, A. (1990). *Arte griego.* C.S.I.C.

Boada, U. (2017). *La senectud del capitalismo. Un reto a la juventud.* EDlibros.

Boardman, J. (1988). *Arte y arquitectura griegos.* In *Historia Oxford del mundo clásico. I. Grecia* (pp. 311–353). Madrid, Alianza.

Bobbio, N. (2007). *El futuro de la democracia.* FCE.

Bobbio, N., Pontara, F., & Veca, S. (1985). *Crisis de la democracia.* Ariel.

Bohm, D., & Peat, F. D. (2007). *Ciencia, orden y creatividad. Las raíces creativas de la ciencia y la vida*. Kairós.

Bohme, G. (2000). *Platons theoretische Philosophie*. Metzler.

Boix, R., & Lazzeretti, L. (2012). Las industrias creativas en España: una panorámica. *Investigaciones Regionales, 22*, 181–206.

Boltanski, L., & Chiapello, É. (2002). *El nuevo espíritu del capitalismo*. Akal.

Boorstin, D. J. (2008). *Los creadores*. Crítica.

Borisonik, H. (2011). Pensando el trabajo a través de Aristóteles. *Astrolabio. Revista internacional de filosofía, 12*, 1–8.

Bowra, C. M. (1983). *La Atenas de Pericles*. Alianza.

Bronowski, J. (1978). *El sentido común de la ciencia*. Península.

Bru, M. (1990). La mujer en el arte prehistórico. Simbología y Representación. *Actas de las Terceras Jornadas de Investigación Interdisciplinar*, 17–28.

Bustamante, E. (1995). De las industrias culturales al entretenimiento. La creatividad, la innovación...viejos y nuevos señuelos para la investigación de la cultura. *Diálogos. Revista Académica de la Federación Latinoamericana de Comunicación*.

Buxton, R. (2000). *El imaginario griego. Los contextos de la mitología*. Cambridge University Press.

Caillois, R. (1986). *Los juegos y los hombres. La máscara y el vértigo*. FCE.

Campbell, J. (1992). *Diosas*. Atalanta.

Campbell, J. (2015). *Las Máscaras de Dios: Mitología Occidental*. Alianza.

Campbell, J. (2019). *Tú eres eso. Las metáforas religiosas y su interpretación*. Atalanta.

Cassirer, E. (1989). *Esencia y efecto del concepto de símbolo*. FCE.

Castoriadis, C. (1998). La polis griega y la creación de la democracia. In *Los dominios del hombre. Las encrucijadas del laberinto* (pp. 97–132). GEDISA.

Castoriadis, C. (2006). *Lo que hace a Grecia. 1. De Homero a Heráclito. Seminarios 1982–1983. La creación humana II*. FCE.

Castoriadis, C. (2008). *El mundo fragmentado*. Caronte.

Castoriadis, C. (2017). Tiempo y creación. In C. Sánchez Capdequí (Ed.), *La creatividad social: narrativas de un concepto actual* (pp. 21–47). CIS.

Cencillo, L. (1970). *Mito. Semántica y realidad*. Biblioteca de Autores Cristianos.

Chantraine, P. (1999). *Dictionnaire étymologique de la langue grecque. Histoire des mots*. Klincksieck.

Chown, M. (2007). *El zoo cuántico. Guía turística del interminable Universo*. La liebre de marzo.

Cioran, E. M. (1986). *La caída en el tiempo*. Planeta Agostini.

Cirlot, J. E. (1988). *Diccionario de símbolos*. Editorial.

Collins, R. (2009). *Cadenas de rituales de interacción*. Anthropos.

Cristiano, J. L. (2009). Imaginario instituyente y teoría de la sociedad. *RES. Revista Española de Sociología, 11*, 101–120.

Cristiano, J. L. (2010). La "Creatividad" de la Acción: la Teoría Joasiana y la cuestión de lo Imaginario. *Nómadas, Revista Crítica de Ciencias Sociales y Jurídicas, 25*, 1–19.

Cristiano, J. L. (2012). La creatividad como aspecto de un replanteamiento de la Teoría Sociológica de la Acción. *Cuadernos de Filosofía Latinoamericana, 33*(106), 33–62.

Damasio, A. (2010). *Y el cerebro creó al hombre: ¿cómo pudo el cerebro generar emociones, sentimientos, ideas y el yo?* Destino.

Danto, A. C. (1999). *Después del fin del arte. El arte contemporáneo y el linde de la historia*. Paidós.

De Carolis, M. (2017). *La paradoja antropológica. Nichos, micromundos, disociación psíquica*. Editorial Quadrata/Iluminuras.

De Coulanges, F. (1986). *La ciudad antigua*. Biblioteca EDAF.

De Souza, B. (n.d.). Globalización y Democracia. *Ponencia presentada en el Foro Social Mundial Temático*.

Delgado, M. (2016). *Ciudadanismo. La reforma ética y estética del Capitalismo*. Catarata.

Delporte, H. (1982). *La imagen de la mujer en el arte prehistórico*. Colegio Universitario de ediciones Istmo.

Dewey, J. (1996). Democracia creativa: la tarea ante nosotros. In *Liberalismo y otros ensayos* (pp. 199–205). Edicions Alfons El Magnànim.

Dodds, E. R. (1983). *Los griegos y lo irracional*. Alianza.

Domínguez, J. (1997). Estética hermenéutica y hermenéutica de la imagen. La articulación de imagen y lenguaje. In H. G. Gadamer & G. Boehm (Eds.), *Estudios de Filosofía* (Vol. 15–16, pp. 91–111).

Donald, M. (2017). Una aproximación evolutiva a la cultura. Consecuencias del estudio de la época axial. In C. Sánchez Capdequí (Ed.), *La creatividad social: narrativas de un concepto actual* (pp. 47–73). CIS.

Dunn, M. (2008). *Diosas. La canción de Eva. El renacimiento del culto a lo femenino*. Malsinet editor, colección Nueva mujer.

Duque, F. (2000). *Filosofía para el fin de los tiempos*. Akal.

Durkheim, É. (1987). *Historia de la educación y de las doctrinas pedagógicas*. La Piqueta.

Durkheim, É. (1993). *Las formas elementales de la vida religiosa*. Alianza.

Durkheim, E. (2017). La efervescencia religiosa. In C. Sánchez Capdequí (Ed.), *La creatividad social: narrativas de un concepto actual* (pp. 151–169). CIS.

Duvignaud, J. (1966). *Sociología del teatro. Ensayo sobre las sombras colectivas*. FCE.

Duvignaud, J. (1980). *Le jeu du jeu*. Bolland, Paris. Mesa redonda. Nuevos horizontes. El juego. p. 118. Retrieved March 13, 2020 from www.libertario.arte.bo/biblioteca/Mesa%20redonda%20Nuevos%20Horizontes%20-%20Jean%20Duvignaud%2C%20el%20juego%20del%20juego.html

Duvignaud, J. (1982). *El juego del juego*. FCE.

Eco, U. (1972). *La definición del arte*. Ediciones Martínez Roca.

Eisenstadt, S. N. (1986a). *The origins and diversity of axial age civilizations*. State University.

Eisenstadt, S. N. (1986b). The axial age breakthroughs-their characteristics and origins. In *The origins and diversity of axial age civilizations* (pp. 1–29). State University of New York.

Eisenstadt, S. N. (2017). La dimensión civilizacional en el análisis sociológico. In C. Sánchez Capdequí (Ed.), *La creatividad social: narrativas de un concepto actual* (pp. 73–95). CIS.

Eliade, M. (1952). *El mito del eterno retorno*. EMECE.

Eliade, M. (1989). *Herreros y alquimistas*. Alianza.

Eliade, M. (2011). *Ocultismo, brujería y modas culturales*. Paidós.

Elias, N. (1991). *Mozart. Sociología de un genio*. Península.

Elias, N. (2000). *Teoría del símbolo. Un ensayo de antropología cultural*. Península.

Elias, N. (2012). *La Soledad de los moribundos*. FCE.

Elias, N. (2013). *Sobre el tiempo*. FCE.

Engels, F. (2013). *El origen de la familia, la propiedad privada y el Estado*. Alianza.

Errejón, I. (2011). El 15-M como discurso contrahegemónico. *Encrucijadas. Revista Crítica de Ciencias Sociales, 2*, 120–145.

Farfán, H. R. (1999). Ni acción ni sistema: el tercer modelo de acción de Hans Joas. *Sociológica*, año 4, número 40, *Perspectivas contemporáneas en la Teoría Social*, mayo–agosto, 35–63.

Farson, R. E. (1973). Las potencialidades humanas. In VVAA (Ed.), *Ensayos sobre el Apocalipsis*. Kairós.

Fernández Galán, C., & García Ramírez, J. (2012). Lectura de la imagen: ¿semiótica o hermenéutica?. *Imaginario Visual Investigación • Arte • Cultura*. Año 2, No. 4. Noviembre–Abril 2013, pp. 48–55.

Fernández, J. L., Parapar, C., & Ruiz M. (2010). El envejecimiento de la población. *Lynchos. Cuadernos de la Fundación General del C.S.I.C.*, No. 2. Edición digital.

Fernández-Armesto, F. (1999). *Historia de la verdad y una guía para perplejos*. Herder.

Fernández-Planells, A., Feixa Pampols, C., & Figueroas-Maz, M. (2013). 15-M En España: Diferencias y Similitudes en las Prácticas Comunicativas con los Movimientos Previos. *Última década, Proyecto Juventudes, 21*(39), 115–138.

Festugiére, A. (1986). *La esencia de la tragedia griega*. Ariel.

Field, F. (1991). *British and French writers in the First World War*. Cambridge University Press.

Figueras, A. J., & Morero, H. A. (2013). La teoría del consumo y de los ciclos en Thorstein Veblen. *Revista de Economía Institucional, 15*(28), 159–182.

Finkelstein, I., & Silberman, N. A. (2005). *La Biblia desenterrada. Una nueva visión arqueológica del antiguo Israel y de los orígenes de sus textos sagrados*. Siglo XX.

Finley, M. I. (1975). *Los griegos de la Antigüedad*. Nueva Colección Labor.

Fischer, E. (1986). *La necesidad del arte*. Planeta-Agostini.

Flores D'Arcais, P. (2013). *¡Democracia! Libertad privada y libertad rebelde*. Galaxia Gutenberg-Círculo de Lectores.

Florida, R. (2009). *Las ciudades creativas. Por qué donde vives puede ser la decisión más importante de tu vida*. Paidós.

Florida, R. (2010). *La Clase Creativa. La transformación de la cultura del trabajo y el ocio en el siglo XXI*. Paidós.

Flynn, T. (2002). *El cuerpo en la escultura*. Akal.

Fraiman, J. A. (2015). Algunas consideraciones sobre el concepto de trabajo en Karl Marx y el análisis crítico de Jürgen Habermas. *Trabajo y Sociedad*, No. 25.

Freud, S. (1908). El poeta y los sueños diurnos. *Obras completas, 4*, 1343–1348.

Freud, S. (1985). *Tótem y Tabú*. Alianza Editorial-Club Internacional del Libro.

Freund, J. (1979). Observaciones sobre dos categorías de la dinámica polemógena. In R. Starn et al. (Eds.), *El Concepto de Crisis*. ediciones Megápolis.

Frye, N. (1988). *El Gran Código*. Gedisa.

Frye, N. (1996). *Poderosas palabras. La Biblia y nuestras metáforas*. Muchnik.

Fumagalli, A. (2010). *Bioeconomía y capitalismo cognitivo. Hacia un nuevo paradigma de acumulación*. Traficantes de Sueños.

Gadamer, H.-G. (1996). *La actualidad de lo bello*. Paidós.

Gadamer, H.-G. (1997). *Mito y razón*. Grupo Planeta (GBS).

Galbraith, J. K. (1984). *La Era de la Incertidumbre*. Plaza & Janés.

Gallego, J. (2003). Comunidad aldeana y sociabilidad campesina en la Grecia Antigua. In J. Gallego (Ed.), *El mundo rural en la Grecia antigua* (pp. 327–380). Akal.
Gamble, C. (2001). *Las sociedades paleolíticas de Europa*. Ariel.
García-Canclini, N. (2005). *La globalización imaginada*. Paidós.
Gardner, H. (2010). *Mentes creativas: una anatomía de la creatividad*. Paidós.
Garnsey, P. (2003). El campesinado: subsistencia y supervivencia. In J. Gallego (Ed.), *El mundo rural en la Grecia antigua* (pp. 71–102). Akal.
Giddens, A. (1993). *Consecuencias de la Modernidad*. Alianza.
Giddens, A., & Luhmann, N. (1996). *Las consecuencias perversas de la modernidad: modernidad, contingencia y riesgo* (Vol. 12). Anthropos Editorial.
Gil, L. (1989). La ideología de la democracia ateniense. In *Cuadernos de Filología Clásica*, No. 23, 30–50. Universidad Complutense.
Gimbutas, M. (1991). *Diosas y dioses de la vieja Europa*. Ediciones Istmo.
Gimbutas, M. (1997). La religiosidad de la Diosa en la Europa Mediterránea. In *Tratado de Antropología de lo sagrado: las civilizaciones del Mediterráneo y lo Sagrado* (pp. 41–62). Trotta.
Glacken, C. J. (1996). *Huellas en la playa de Rodas. Naturaleza y cultura en el pensamiento occidental desde la Antigüedad hasta finales del siglo XVIII*. Ediciones del Serbal.
Glotz, G. (1957). *La ciudad griega*. Unión tipográfica Hispano Americana.
González, J. M. (2006). *La diosa Fortuna. Metamorfosis de una metáfora política*. Antonio Machado libros.
González García, J. M. (1992). *Las huellas de Fausto. La herencia de Goethe en la sociología de Max Weber*. Tecnos.
González García, J. M. (1998). Sociología e iconología. *Madrid: REIS, 84*, 23–43.
Granés, C. (2011). *El puño invisible. Arte, revolución y un siglo de cambios culturales*. Taurus.
Gratton, L. (2012). *Prepárate: el futuro del trabajo ya está aquí*. Galaxia Gutenberg-Círculo de Lectores.
Green, A. (1993). La madre muerta. In *Narcisismo de Vida, narcisismo de Muerte*. Ed. Amorrortu.
Greene, B. (2006). *El universo elegante. Supercuerdas, dimensiones ocultas y la búsqueda de una teoría final*. Drakontos bolsillo.
Gregori, E. (2002). Individuo y naturaleza en Schiller. La noción de juego. *Materia, 2, Naturaleses*, 35–55.

Grondin, J. (2014). *A la escucha del sentido: conversaciones con Marc-Antoine Vallée*. Herder.
Guasch, A. M. (2002). *El arte último del siglo XX. Del posminimalismo a lo multicultural*. Alianza.
Guellouz, A. (1995). El islam. In J. Delumeau (Ed.), *El hecho religioso. Enciclopedia de las grandes religiones* (pp. 253–331). Alianza.
Guilaine, J., & Zammit, J. (2002). *El camino de la guerra. La violencia en la prehistoria*. Ariel.
Habermas, J. (1984). *Ciencia y Técnica como ideología*. Tecnos.
Habermas, J. (1989). *El discurso filosófico de la modernidad*. Taurus.
Halbwachs, M. (2011). *La memoria colectiva*. Miño y Dávila.
Hanna, M. (1996). *The mobilization of intellect. French scholars and writers during the Great War*. Harvard University Press.
Hauser, A. (1988). *Historia social del arte y de la literatura*. Labor.
Hawkins, S. (2005). *El universo en una cáscara de nuez*. Crítica/Planeta.
Hayoun, M. R. (1995). El judaísmo. In J. Delumeau (Ed.), *El hecho religioso. Enciclopedia de las grandes religiones* (pp. 201–252). Alianza.
Heidegger, M. (1987). *De camino al habla*. Delós.
Heidegger, M. (2003). *Ser y tiempo*. Trotta.
Heidegger, M. (2016). *El origen de la obra de arte*. La Oficina.
Held, D. (1993). *Modelos de democracia*. Alianza.
Hidalgo González, J. G. (2001). *El envejecimiento. Aspectos sociales*. Editorial de la Universidad de Costa Rica.
Hobsbawm, E. (2001). *Historia del Siglo XX*. Crítica.
Hobsbawm, E. (2013). *Un tiempo de rupturas. Sociedad y cultura en el siglo XX*. Crítica.
Homero. (2001). *Himnos homéricos* (J. García, Introducción general y A. Bernabé, traducción y notas). Clásica Gredos.
Hudson, W. (1982). *The Marxist philosophy of Ernst Bloch*. Springer.
Huizinga, J. (1987). *Homo Ludens*. Alianza.
Hulin, M., & Kapani, L. (1995). El hinduísmo. In J. Delumeau (Ed.), *El hecho religioso. Enciclopedia de las grandes religiones* (pp. 333–402). Alianza.
Husain, S. (2001). *La Diosa. Creación, fertilidad y abundancia. Mitos y arquetipos femeninos*. Taschen.
Hynes, S. (1992). *A war imagined. The First World War and English culture*. Pimlico.
Illouz, E. (2009). *El consumo de la utopía romántica. El amor y las contradicciones culturales del capitalismo*. Katz.

Ingham, G. (2010). *Capitalismo*. Alianza.

Inglehart, R. (1991). *El cambio cultural en las sociedades industriales avanzadas*. CIS.

Innerarity, D. (2004). *La sociedad invisible*. Espasa Calpe.

Iriarte, A. (2010). Atenas o el arte de lo político. *Revista Bitarte, 50*, 29–42.

Iriarte, A., & González, M. (2008). *Entre Ares y Afrodita. Violencia del erotismo y erótica de la violencia en la Grecia Antigua*. Abada editores.

Isnenghi, M. (1989). *Il Mito della Grande Guerra*. Il Mulino.

Jaspers, K. (2017). *Origen y meta de la historia*. Acantilado.

Joas, H. (2002). *Creatividad, Acción y Valores*. Universidad Autónoma Metropolitana.

Joas, H. (2013). *La creatividad de la acción*. CIS.

Jung, C. G. (2014). *Sobre el fenómeno del espíritu en el arte y en la ciencia*. Trotta.

Jung, C. G. (2015). *Arquetipos e inconsciente colectivo*. Paidós.

Jünger, E. (1998). *El libro del reloj de arena*. Tusquets.

Kant, I. (2001). *Crítica del juicio*. Austral.

Kerényi, K. (2010). *Imágenes primigenias de la religión griega. II. Hermes el conductor de almas. El mitologema del origen de la vida masculina*. editorial Sexto Piso.

Kitto, H. D. F. (1979). *Los griegos*. EUDEBA.

Knöbl, W. (2017). El debate de 'las modernidades múltiples, su fundamento sociológico-religioso y sus escollos. In C. Sánchez Capdequí (Ed.), *La creatividad social: narrativas de un concepto actual* (pp. 263–315). CIS.

Koselleck, R. (2016). *historia/Historia*. Trotta.

Krauss, R. (1996). *La originalidad de la Vanguardia y otros mitos modernos*. Alianza.

Lamo de Espinosa, E. (1990). *La sociedad reflexiva*. Siglo XXI.

Lessing, E., & Sollers, P. (1994). *Mujeres, mitologías*. M Moleiro editor.

Lévêque, P. (1978). Las diferenciaciones sociales en el seno de la democracia ateniense del siglo V. In D. Roche (Ed.), *Órdenes, estamentos y clases*. Siglo XXI.

Lévêque, P. (1993). El papel de la religión en la génesis de las ciudades. *Revista de Occidente, 143*, 43–60.

Luhmann, N. (2017). La autopoiesis de los sistemas sociales. In C. Sánchez Capdequí (Ed.), *La creatividad social: narrativas de un concepto actual* (pp. 229–249). CIS.

Lince, R. (2009). *Hermenéutica, arte y ciencia de la interpretación*. UNAM.

Lizcano, E. (2003). Imaginario colectivo y análisis metafórico. transcripción de la conferencia inaugural del *Primer Congreso Internacional de Estudios sobre*

Imaginario y Horizontes Culturales que se celebró en la Universidad Autónoma del Estado de Morelos, Cuernavaca, México, del 6 al 9 de mayo. Retrieved October 7, 2019, from http://www.unavarra.es/puresoc/pdfs/c_salaconfe/SC-Lizcano-2.pdf
Llanos Melussa, E. (2018). La creatividad, ¿qué es y cómo podemos estimularla? *Creces*, 6–14.
Longair, M. S. (1998). *La evolución de nuestro universo*. Cambridge University Press.
López Aguilar, J. F. (2011). El hartazgo de las instituciones. *Claves de razón práctica, 215*, 20–28.
López Mondéjar, L. (2015). *Una espina en la carne. Psicoanálisis y creatividad*. Psimática Clínica.
Lucie-Smith, E. (1994). *La sexualidad en el arte occidental*. Destino.
Macpherson, C. B. (1981). *La Democracia Liberal y su época*. Alianza.
Maffesoli, M. (2004). *El tiempo de las tribus. El ocaso del individualismo en las sociedades posmodernas*. Siglo XXI.
Maffesoli, M. (2020). De L'Universalisme à l'uni-diversité postmoderne. In C. Sánchez Capdequí & J. A. Roche Cárcel (Eds.), *Modernidades regresivas*. CIS.
Mannheim, K. (1993). El problema de las generaciones. *REIS, 62*, 193–242.
Mannheim, K. (2018). *El hombre y la sociedad en la época de crisis*. CIS.
Marangudakis, M. (2016). Visions of brotherhood; a comparative analysis of direct democracy in ancient and modern Greece. *Política y Sociedad, 53*(3), 773.
Marramao, G. (1989). *Poder y secularización*. Península.
Martin, M. (2008). La teoría de las generaciones de Ortega y Gasset: una lectura del siglo XXI. *Tiempo y espacio.*, Year 17, *20*, 98–110.
Marx, K. (1985). *Trabajo asalariado y capital*. Planeta-Agostini.
Maté, C. (2004). El juego en los antropoides y los niños. *Estudios de Psicología, 25*(2), 149–168.
Mead, G. H. (1938). *The philosophy of the act*. The University of Chicago Press.
Mead, G. H. (1991). *Espíritu, Persona y Sociedad. Persona y Sociedad desde el punto de vista del conductismo social*. Paidós.
Mills, C. W. (1999). *La imaginación sociológica*. FCE.
Mitchell, W. J. T. (2019). *La ciencia de la imagen. Iconología, Cultura Visual y estética de los medios*. Akal.
Moliner, M. (2007). *Diccionario de uso del español Maria Moliner*. Gredos.

Monforte, J. M. (n.d.). *La Biblia y el sentido del Universo*. Biblioteca Católica Digital. Retrieved January 30, 2018, from www.almudi.org

Monod, J. (2007). *El azar y la necesidad*. Tusquets.

Montesquieu. (1997). *Cartas persas*. Cátedra.

Morin, E. (1979). Para una crisología. In R. Starn et al. (Eds.), *El Concepto de Crisis*. ediciones Megápolis.

Morin, E. (1981). *El método I. La naturaleza de la naturaleza*. Cátedra.

Morin, E. (1984). *Ciencia con consciencia*. Anthropos.

Morin, E. (1994). *El hombre y la muerte*. Kairós.

Morin, E. (1998). Epistemología de la complejidad. In D. F. Schnitman (Ed.), *Nuevos paradigmas, cultura y subjetividad* (pp. 421–446). Paidós.

Moscone, R. O. (2012). El miedo y sus metamorfosis. *Psicoanálisis, XXIV*(1), 53–78.

Mossé, C. (1987). *Historia de una democracia: Atenas*. Akal.

Mossé, C. (1993). El hombre y la economía. In J.-P. Vernant (Ed.), *El hombre griego*. Alianza.

Mossé, C. (2001). *La mujer en la Grecia clásica*. Nerea.

Mumford, L. (1979). *La ciudad en la historia*. Infinito.

Muñoz, I. (1975). La ideología de la aristocracia griega antigua. *Revista de Estudios Políticos, 199*, 65–82.

Muñoz, B. (2001). Reflexiones sobre la sociología de la cultura y de la música en la obra de Max Weber: un análisis crítico. *Sociedad y Utopía. Revista de Ciencias Sociales, 18*, 23–37.

Muñoz, B. (2015). *La dominación simbólica en la globalización. Una Teoría Crítica sobre la Postmodernidad*. Fundamentos.

Nair, S., & de Lucas, J. (1999). *Inmigrantes. El desplazamiento en el mundo*. Ministerio de Trabajo y Asuntos Sociales. Secretaría General de Asuntos Sociales. Instituto de Migraciones y Servicios Sociales.

Naredo, J. M. (2002). Configuración y crisis del mito del trabajo. *Scripta Nova: Revista electrónica de geografía y ciencias sociales*, Extra Issue No. 6, 119.

Neumann, E. (2004). El hombre creador y la Transformación. In E. Neumann, M. Eliade, G. Durand, H. Kawai, & V. Zuckerhand (Eds.), *Los dioses ocultos: Círculo de Eranos II*. Anthropos.

Nisbet, R. (1991). *Historia de La idea del progreso*. Editorial Gedisa, SA.

Nocera, P. (2009). Los usos del concepto de efervescencia y la dinámica de las representaciones colectivas en la sociología durkheimiana. *REIS, 127*, 93–119.

Noya, J. (2010). Sociología de la creatividad. In J. Noya, F. Del Val, & M. Pérez Colman (Comps), *Musyca. Música, sociedad y creatividad artística.* Biblioteca Nueva.

Nussbaum, M. C. (1995). *La fragilidad del bien. Fortuna y ética en la tragedia y la filosofía griega.* Visor.

Ortega y Gasset, J. (1983). *El tema de nuestro tiempo.* Alianza Editorial/Revista de Occidente.

Ortega y Gasset, J. (2008). *En torno a Galileo.* Revista de Occidente-Alianza.

Ortiz-Osés, A. (1993). *Las claves simbólicas de nuestra cultura: Matriarcalismo, patriarcalismo y fratriarcalismo.* Anthropos.

Parménides. (2007). *Poema.* Ediciones Folio.

Parsons, T. (1967). Evolutionary universals in society. In *Sociological theory and modern society.* The Free Press.

Pastor Verdú, J. (2011). El Movimiento 15 M. Un nuevo actor sociopolítico. *Sociedad y Utopía. Revista de Ciencias Sociales, 38*, 109–124.

Pérez, J., & Abellán, A. (2010). Sociedades longevas: un desafío para el Siglo XXI. *Lynchon. Cuadernos de la Fundación General CSIC.* Digital Edition No. 2.

Pérez Díaz, J. (1998). La demografía y el envejecimiento de las poblaciones. In A. S. Staab & L. C. Hodges (Eds.), *Enfermería Gerontológica* (pp. 451–463). McGraw Hill.

Pérez Díaz, J. (2016). El temor al envejecimiento demográfico. In J. S. Humet et al. (Eds.), *Edades en transición. Envejecer en el siglo XXI* (pp. 44–54). Ariel.

Picó, J. (1988). Introducción. *Modernidad y Posmodernidad.* Alianza.

Platón. (2000). *Diálogos*, vol. IV *(República)*, Introducción, traducción y notas de Conrado Eggers Lan). Editorial Gredos.

Platón, D. (2003). *Obra completa* (Vol. 9 vol). Editorial Gredos.

Polanyi, K. (2017). *La gran transformación. Los orígenes políticos y económicos de nuestro tiempo.* FCE.

Pope, R. (2005). *Creativity. Theory, history, practice.* Routledge.

Prochasson, C. (1993). *Les Intellectueles, le socialisme et la guerre, 1900–1938.* Seuil.

Prochasson, C., & Rasmussen, A. (1996). *Au nom de la patrie. Les intellectuels et la Première guerre mondiale, 1910–1919.* La Découverte.

Ramos Torre, R. (1999). *Red, identidad, espacio y tiempo* (p. 86). *Revista Española de Investigaciones Sociológicas.*

Ramos Torre, R. (2012). Los avatares de lo trágico en la sociología: de Tocqueville a Luhmann. In J. A. Roche Cárcel (Ed.), *La Sociología como una de las Bellas Artes. La influencia de la literatura y de las artes en el pensamiento sociológico.* Anthropos.

Reckwitz, A., & Sánchez Capdequí, C. (2017). La estetización y el dispositivo de la creatividad: el régimen social de la novedad estética. In *La creatividad social: narrativas de un concepto actual* (pp. 347–374). Centro de Investigaciones Sociológicas (CIS).

Requejo Coll, F. (1990). *Las democracias. Democracia antigua, democracia liberal y Estado de Bienestar*. Ariel.

Ricoeur, P. (1967). *The symbolism of evil*. Harper and Row.

Ricoeur, P. (1988a). *Hermenéutica y Acción. De la Hermenéutica del Texto a la Hermenéutica de la Acción*. Prometeo Libros-UCA.

Ricoeur, P. (1988b). *Time and narrative* (Vol. III, K. Blarney & D. Pellauer, Trans.). Chicago University Press.

Ricoeur, P. (2010). *La memoria, la historia, el olvido*. Trotta.

Ries, J. (1997). *Tratado de antropología de lo sagrado: las civilizaciones del Mediterráneo y lo sagrado*. Trotta.

Rifkin, J. (2000). *La era del acceso. La revolución de la nueva economía*. Paidós.

Rivero, C. I. (2002). El aporte de Edgar Morín al pensamiento social contemporáneo, desde una epistemología de la complejidad. *Salud de los Trabajadores*/ Volumen 10, Edición Especial No. 1 & 2.

Robert, J.-N. (1995). El budismo. In J. Delumeau (Ed.), *El hecho religioso. Enciclopedia de las grandes religiones* (pp. 427–474). Alianza.

Roche Cárcel, J. A. (2007). De la nostalgia por el origen a la pasión por el final. De la génesis de la sociedad al fin de lo social. In J. A. Roche Cárcel (Ed.), *Espacios y tiempos inciertos de la cultura* (pp. 37–68). Anthropos.

Roche Cárcel, J. A. (2012). Tiempo líquido y cultura de la incertidumbre. *International Review of Sociology, 22*(1), 137–162.

Roche Cárcel, J. A. (2013a). La frágil construcción de la democracia en la Grecia Antigua. In the monograph *Transiciones. La fragilidad de la democracia*, issue coordinated by J. A. Roche Cárcel, *Res Publica. Revista de filosofía política*, No. 30, Year 16, 15–58.

Roche Cárcel, J. A. (2013b). El incierto sentido del trabajo en la sociedad de riesgo global. Barataria: revista castellano-manchega de ciencias sociales, ISSN 1575-0825, No. 15, 207–226.

Roche Cárcel, J. A. (2017). *Entre el Monte de Apolo y la vid de Dioniso. Naturaleza, Dioses y Sociedad en la arquitectura teatral de la Grecia Antigua*. Anthropos.

Rodríguez Adrados, F. (1983). *La Democracia ateniense*. Alianza.

Rodríguez Ferrándiz, R. (2011). De industrias culturales a industrias del ocio y creativas: los límites del "campo" cultural. *Revista Comunicar*, 36, *La televisión y sus nuevas expresiones, 18*, 149–156.

Rovelli, C. (2018). *El orden del tiempo*. Anagrama.

Rubio Arostegui, J. A., Pecourt, J., & Rius Ulldemolins, J. (2016). Usos y abusos de la creatividad. Sociología de los procesos creativos, transiciones a lo digital y políticas creativas. *Debats, 130*(2), 125–145.

Sabariz, C. R. (2018). Una salida al dualismo: Juego y trabajo en John Dewey. *Éndoxa, 41*, 135–156.

Sánchez Belando, M. V., Rius Ulldemolins, J., & Zarlenga, M. I. (2012). ¿Ciudad creativa y ciudad sostenible: un análisis crítico del "modelo Barcelona" de políticas culturales. *Revista Crítica de Ciências Sociais, 99*, 31–50.

Sánchez Capdequí, C. (2004). *Las máscaras del dinero. El simbolismo social de la riqueza*. Anthropos-Universidad Autónoma Metropolitana.

Sánchez Capdequí, C. (2010). *En los límites de la con-fusión. Miedos, riesgos y urgencias de la Sociedad de la Información*. Libros de la Catarata.

Sánchez Capdequí, C. (2014). *El dinamismo de los valores. Crisis y creatividad en la sociedad moderna*. Anthropos.

Sánchez Capdequí, C. (2018). Las sociologías post 68 y la creatividad social. *Arbor, 194*(787), a435.

Sánchez Ron, J. M. (2000). *El siglo de la Ciencia*. Taurus.

Sanz, J., & Mateos, Ó. (2011). 15-M. Apuntes para el análisis de un movimiento en construcción. *Revista de Fomento Social, 66*, 517–544.

Sassen, S. (2007). *Una sociología de la globalización*. Katz.

Schiller, F. (1985). *Sobre poesía ingenua y poesía sentimental y una polémica Kant, Schiller, Goethe, Hegel*. Icaria.

Schiller, F. (1990). *Kallias; Cartas sobre la educación estética del hombre*. Anthropos.

Schipper, K. (1995). El taoísmo. In J. Delumeau (Ed.), *El hecho religioso. Enciclopedia de las grandes religiones* (pp. 509–548). Alianza.

Schopenhauer, A. (2017). *El mundo como voluntad y representación*. Alianza.

Schrödinger, E. (2017). *¿Qué es la vida?* Tusquets.

Seco, M. (1999). *Diccionario del español actual*. Aguilar.

Sennett, R. (2000). *La corrosión del carácter. Las consecuencias personales del trabajo en el nuevo capitalismo*. Anagrama.

Serrano, C. (2014). *Sociología de la economía creativa en Aragón. Números, trayectorias vitales y entornos*. Doctoral Thesis, Universidad de Zaragoza, Zaragoza.

Shattuck, R. (1991). *La época de los banquetes. Orígenes de la vanguardia en Francia: de 1885 a la Primera Guerra Mundial*. Visor.

Simmel, G. (1986). *El individuo y la libertad*. Península.

Simmel, G. (2014). *Intuición de la vida*. Prometeo.

Simmel, G. (2017). La trascendencia de la vida. In C. Sánchez Capdequí (Ed.), *La creatividad social: narrativas de un concepto actual* (pp. 213–229). CIS.

Sloterdijk, P. (2019). *En el mundo interior del capital. Para una teoría filosófica de la globalización*. Siruela.

Smith, P. (2003). *Arte rupestre paleolítico en Cantabria*. Asociación Cántabra para la defensa del patrimonio Subterráneo.

Smith, A. (2007). *La riqueza de las naciones*. Alianza editorial.

Sombart, W. (2009). *Lujo y capitalismo. Economía de la ostentación*. Sequitur.

Soros, G. (1999). *Crisis del Capitalismo Global*. La. Sudamericana.

Starn, R. (1979). Metamorfosis de una noción. Los historiadores y la "crisis". In R. Starn et al. (Eds.), *El Concepto de Crisis*. ediciones Megápolis.

Steiner, G. (2002). *Gramáticas de la creación*. Siruela.

Stromberg, R. (1982). *Redemption by war: The intellectuals and 1914*. The Regents Press of Kansas.

Subirats, E. (1989). *El final de las vanguardias*. Anthropos.

Tatarkiewicz, W. (1987). *Historia de la Estética. I. La estética antigua*. Akal.

Tezanos, J. F. (2001). *El trabajo perdido. ¿Hacia una civilización postlaboral?* Biblioteca Nueva.

Thaa, W. (2008). Democracia y crítica de la civilización en Max Weber y Hannah Arendt. *Revista española de ciencia política, 19*(October), 9–40.

Thiebaut, C. (1996). La mal llamada postmodernidad (o las contradanzas de lo moderno). In V. Bozal (Ed.), *Historia de las ideas estéticas y de las teorías artísticas contemporáneas* (Vol. II). Visor.

Thom, R. (1979). Crisis y catástrofe. In R. Starn et al. (Eds.), *El Concepto de Crisis*. ediciones Megápolis.

Tilly, C. (2010). *Democracia*. Akal.

Todorov, T. (2012). *Los enemigos íntimos de la democracia*. Galaxia Gutenberg-Círculo de Lectores.

Tönnies, F. (2011). *Comunidad y asociación. El comunismo y el socialismo como formas de vida social*. Biblioteca Nueva-Minerva.

Touraine, A. (1995). *Producción de la Sociedad*. Instituto de Investigaciones Sociales UNAM-IFAL.

Trías, E. (1984). *Drama e identidad*. Ariel.

Trías, E. (2019). *La filosofía y su sombra*. Galaxia Gutenberg.

Valencia López, H. (2016). La política y la democracia como creaciones imaginarias: de los griegos a nosotros. *Atenea* 513, I Sem., 125–135.

Vallespín, F. (2011). La fatiga democrática. *Claves de razón práctica, 215*, 10–18.

Vandermeersch, L. (1995). El confucionismo. In J. Delumeau (Ed.), *El hecho religioso. Enciclopedia de las grandes religiones* (pp. 333–402). Alianza.
Vattimo, G. (1995). *El fin de la modernidad. Nihilismo y hermenéutica en la cultura posmoderna*. GEDISA.
Veblen, T. (2008). *Teoría de la clase ociosa*. Alianza.
Vernant, J.-P. (2001). *El individuo, la muerte y el amor en la Antigua Grecia*. Paidós.
Vernant, J.-P. (2003). *El universo, los dioses, los hombres. El relato de los mitos griegos*. Anagrama.
Veyne, P. (1992). El Imperio Romano: Los esclavos; Trabajo y descanso. In P. Ariès & G. Duby (dirs.), *Historia de la vida privada. Imperio Romano y Antigüedad tardía* (Vol. 1). Taurus.
Wallerstein, I. (1997). Incertidumbre y creatividad. *Espacios*, pp. 69–71.
Watson, P. (2002). *Historia intelectual del siglo XX*. Crítica.
Weber, M. (1971). *Sobre la teoría de las ciencias sociales*. Península.
Weber, M. (1984). *Economía y Sociedad. Esbozo de sociología comprensiva*. FCE.
Weber, M. (1997). *Sociología de la religión*. Fondo de Cultura Económica.
Weber, M. (2006). *Conceptos sociológicos fundamentales*. Alianza.
Weber, M. (2007). *Sociología del poder. Los tipos de dominación*. Alianza.
Weber, M. (2017). *Sociología de la religión*. AKAL.
Weisberg, R. W. (1989). *Creatividad. El genio y otros mitos*. Editorial Labor.
West, M. L. (1966). *Hesiod Theogony*. Oxford University Press.
Wilson, E. O. (2018). *Los orígenes de la creatividad humana*. Crítica.
Winnicott, D. W. (2013). *Realidad y juego*. GEDISA.
Wohl, R. (1979). *The generation of 1914*. Harvard University Press.
Ziemann, B., & Hofrichter, A. A. (2016). La violencia como objeto de estudio en las investigaciones recientes sobre la primera guerra mundial. *Historia Social, 84*, 141–159.
Zweig, S. (2015). *El misterio de la creación artística*. Sequitur.

Index[1]

[1] Note: Page numbers followed by 'n' refer to notes.

J. A. Roche Cárcel, *Creativity and Time: A Sociological Exploration*, Palgrave Studies in Creativity and Culture, https://doi.org/10.1007/978-3-030-84838-5

The manufacturer's authorised representative in the EU is Springer Nature Customer Service Centre GmbH, Europaplatz 3, 69115 Heidelberg, Germany. If you have any concerns regarding our products, please contact ProductSafety@springernature.com

Printed and bound by CPI Group (UK) Ltd, Croydon, CR0 4YY
22/09/2026
02221066-0002